The

MISSION
OF A
LIFETIME

The

MISSION

OF A

LIFETIME

LESSONS FROM THE MEN
WHO WENT TO THE MOON

BASIL HERO

GRAND CENTRAL
PUBLISHING

NEW YORK   BOSTON

Grand Central Publishing
Hachette Book Group
1290 Avenue of the Americas, New York, NY 10104
grandcentralpublishing.com
twitter.com/grandcentralpub

First Edition: April 2019

Grand Central Publishing is a division of Hachette Book Group, Inc. The Grand Central Publishing name and logo is a trademark of Hachette Book Group, Inc.

The publisher is not responsible for websites (or their content) that are not owned by the publisher.

The Hachette Speakers Bureau provides a wide range of authors for speaking events. To find out more, go to www.hachettespeakersbureau .com or call (866) 376-6591.

Except where otherwise indicated, all photos were provided courtesy of NASA. Used by permission.

LCCN: 2018964674

ISBNs: 978-1-5387-4851-0 (hardcover), 978-1-5387-4850-3 (ebook), 978-1-5387-3406-3 (B&N signed)

Printed in the United States of America

LSC-C

10 9 8 7 6 5 4 3 2 1

In memory of my mother and father,
Angela and Byron

And to those who hold my heart: my loving wife, Marianne,
and my precious daughters, Lexie and Ariane

Contents

FINDING THE EAGLES

Bill Anders still flies his collection of vintage airplanes, 2018
(Courtesy of Basil Hero)

Space is making a comeback. Cable news networks are once again broadcasting the live countdowns of rocket launches—this time of the privately built rockets of billionaire visionaries such as Jeff Bezos and Elon Musk. Their plans for colonizing the moon and Mars are making regular headlines, and a new generation raised on movies like *Gravity* and *The Martian* has gotten the bug for space. In 2017, 18,300 people applied to NASA for fourteen available astronaut slots . . . all of them hopeful that they might, one day, run their gloved fingers over the surface of another world.

How extraordinary to think that with all the technological progress of the last fifty years, no country today has a rocket powerful enough to achieve the six Apollo moon landings that took place between 1969 and 1972. No human has left low earth orbit or kicked up any moon dust since then. That's a long time. What, I wondered, do the twelve of the remaining twenty-four Apollo astronauts who either orbited or walked on the moon think about all of this—not just about the state of today's space program and its future but, more intriguingly, the life lessons they learned from humanity's ultimate road trip of exploration? While their lunar missions have been written about exhaustively, including their autobiographies, those were written years ago and stopped short of probing the more primal questions each of us has wrestled with. It's possible, I thought, that as they were closing in on their nineties, they might be willing (even keen, perhaps) to offer the kind of life reflections generated by the accumulated wisdom of their emeritus years. Reflections about courage. About conquering one's fears. About what constitutes a worthwhile personal and professional life. About the latest information on our home planet's perilous state and their thoughts on the bold plans of innovators like Jeff Bezos to colonize the universe. I decided to make it my mission to hear what they had to say.

The men who went to the moon remain history's most elite fraternity. Their extraterrestrial view of Earth from the moon changed them and the world. As of this writing, only twelve of the twenty-four lunar voyagers are alive. Time was running out to talk to them. The bigger challenge I faced was how to reach them.

The improbable answer came on the night of October 29, 2014, with the arrival of Hurricane Sandy, which mutated into a monster that meteorologists needed a new name to describe. They called it FRANKENSTORM! Lower Manhattan was deluged by a fourteen-foot wall of surging ocean water and went dark. In Connecticut, where I lived, uprooted trees and downed power lines cut the electricity to my Westport home. My immediate concern was for my retired neighbor who moved next door to me the week before. I was unaware that he was one of the nation's leading space historians. When I knocked on his door to check on him, I could see right away that he needed more than candlelight to safely navigate his living room, so I returned with a few spare storm lanterns. The room lit up to reveal the lair of a writer. The center of Bill's living room was dominated by an eight-foot-long table covered with multiple rows of neatly labeled manila file folders arrayed against open books loaded with sticky notes. Clipped newspaper articles were scattered near his computer, and bookcases gobbled up the rest of the room. Above his small, round wooden dining table hung a corkboard with a single sheet of bold lettering spelling out WE ARE THE SPACE RACE.

"Are you the same Bill Burrows," I asked him, "who wrote *This New Ocean: The Story of the First Space Age?*" His big smile said yes. My new neighbor was a Pulitzer Prize finalist, former *New York Times* reporter, retired professor of journalism at NYU, and the author of twelve books (most on space and national security issues).

Bill and I would share our passion for the moon shots as he worked on his newest book, *The Asteroid Threat*, which was published two years after we met. Bill (now eighty-one) became more than a friend as he began advising me in this undertaking to give readers a prescription for living boldly (and deeper reflection) from the men who went to the moon.

The ex-journalist in me kicked in to start tracking down the men I've come to call the Eagles. Bill, of course, was the obvious starting point. But the only contact information he had was for shuttle astronaut Tom Jones (TJ), who wrote *Sky Walking* and gave a dust jacket testimonial for *The Asteroid Threat*. But even TJ had no e-mail addresses for the Eagles, which I soon discovered are known only to a select few who essentially serve as their gatekeepers.

So, who are the gatekeepers? I wondered. TJ suggested going to Andy Turnage, the executive director of the Association of Space Explorers (ASE), to whom he made an introduction. I knew I had one shot with Andy, and condensed my e-mail to him into a few compelling paragraphs about the theme of the book, and gave him a short list of the Eagles I wanted to interview first (among them Apollo 8's Bill Anders, who took the celebrated *Earthrise* photo, which gave humanity its first clear look at Earth from the moon, and ultimately helped launch the environmental movement).

Within the hour, Andy responded to my note. I had passed the first test. He liked what he read, and said he would forward my book proposal to Bill Anders and Charlie Duke, the tenth man to walk on the moon. He did not, however, have contact info for the others I had requested.

My query hit its mark faster than I expected. That same day I received the following response:

Dear Mr. Hero,

I'm Dr. Dydia DeLyser, Archivist and Project Manager for General and Mrs. Anders. As you can see, your message found its way to me. I'd be pleased to learn more about your project and plans. If you'd like to speak by phone rather than email, I'm sure we can set that up.

Best,
Dydia

Dr. DeLyser's note triggered a deep breath of festive relief. But ahead, I knew, lay the make-or-break part. While Anders was intrigued by my e-mail, he understandably wanted his gatekeeper to start vetting me. Anders, my research showed, is a versatile maverick, a bit of a rebel with a scintillating intelligence, and the only astronaut to figure out how to beat one of NASA's intelligence tests. He has a gift for reducing complex issues to their binary essence and has little tolerance for anything less than excellence. Warren Buffett told me that "every move Bill Anders made was smart" when he was CEO of General Dynamics, and it was the reason he gave Anders his board proxy in what is considered one of the great corporate turnarounds in modern business history.

Anders deals in facts and due diligence. Dr. DeLyser's job was to find out if I was worth talking to. My job was to convince her that I was. Google searches can give you the facts of a person but rarely a sense of who they are. Before getting on the phone with Dr. DeLyser, I wanted to know more about her. She is an assistant professor in the Department of Geography and the Environment at California State University, Fullerton. She is not your typical academic. As I read through her faculty bio section, it was this eye-catching notation, under her list of research interests, that told me I was dealing with a creative thinker: "How women pilots in the late 1920s and early 1930s used their practices of flying to advance feminism in the post-suffrage era."

I could see why Anders picked her to organize the history of his life. In a disquieting aside, she recounted the deceptive lengths people have engaged in over the years to get autographs from the Apollo astronauts...including bogus letters from adults pretending to be children dying of cancer, and other sympathy-grabbing devices to get the astronauts' attention and signatures.

I had no idea how bad it was. Obsessive autograph seekers and stalkers have driven the Eagles (the moon walkers, in particular) to closely guard their home addresses and e-mail information. "People want a piece of them," Dr. DeLyser told me. Anything to be able to say they shook the hand, or got the autograph, of an astronaut who went to the moon.

Dr. DeLyser liked the fresh approach I was taking with the Eagles, and asked for my résumé, along with a two-page summary of the book to present to General Anders. I sent it off and hoped I would clear the final hurdle.

To my everlasting gratitude, Anders gave me the opportunity to interview him and invited me to his elegant home overlooking Burrows Bay in Anacortes, Washington. I have interviewed presidents and many other powerful figures at length, but I have never prepared more thoroughly than I did for Anders. When I arrived at the gate, he greeted me in his gardening clothes and semi-joked, as we walked to the house, that he's preparing handicap ramps for the inevitable day when walking will become difficult for him and his wife.

Anders, at eighty-five, is nowhere near that point. What struck me first were his blue eyes, which still have the vigor, watchfulness, and intensity of the ex-fighter-pilot in him. In fact, he says, he's never really gotten out of the cockpit. Like the rest of his lunar brethren who rode the Saturn V, the mightiest rocket ever built, the need for adrenaline-pumping adventure still beats strongly in his octogenarian heart. He sniffs at the idea of playing golf. Not surprisingly, Anders, who is still thin like all the Eagles, says his daily exercise routine is to fly at least one hour a day in one of the vintage World War II planes his wealth has afforded him. Flying keeps him young.

Similar eclectic tastes in history kicked off our conversation, and we moved quickly to the real-time impact his lunar voyage had on his notion of man's place in the universe, the anatomy of courage, and the existence of God. Fully engaged by our talk, he invited me to dinner with two close friends. But before heading out, he couldn't resist having a little fun with me. In his study is his prized collection of meteorites and fossils. "You know," he said, "geologists say you can sometimes tell the origin of a rock by tasting it." I was sensing a setup.

"Run your tongue over this one," Anders said, "and see if you can identify it." I promptly gave it a lick. "You know what it is? Fossilized dinosaur shit."

We both had an uproarious laugh and a wonderful interview.

There is no greater validation than one lunar astronaut recommending you to another, and that's what Anders did when he promised to connect me with Frank Borman, his commander on Apollo 8 and his lifelong friend. Good to his word, I got the following witty e-mail from Bill a few days later on Easter Sunday:

Basil,

The Bunny is bringing you a nice egg this morning. I talked to Frank Borman yesterday and he has agreed to chat with you about your book in Billings, MT. His wife is quite ill and he spends a lot of time with her daily so there will be some comings & goings. You can contact him at the above email or his cell phone.

The same insightful exploration continued with the ninety-one-year-old Borman who, like Bill, still flies his own vintage plane. Borman is as dynamic as ever. In coordinating our interview, he told me he was an early riser and offered to pick me up at my hotel at 8:30 in the morning after finishing his daily workout routine. His day starts at 5:30 a.m. and consists of one hour of weight lifting and treadmill work. From my research of the man, I had a feeling that the West Point graduate, a human dynamo guided by the Academy's motto: *Duty, Honor, Country,* might arrive early.

Acting on my hunch, I decided to take an early walk around the hotel grounds. Within five minutes of inhaling Montana's Big Sky air, I spotted a small SUV pulling into the driveway. A flash of certainty hit me that it was the colonel behind the wheel. He lowered his window, and as I recognized him, the first thing I noticed were his eyes bursting with alertness, the same eyes whose deep space view of the Earth from the moon rewired his brain's perception of our home planet. He was clearly pleased that I was waiting for him ahead of schedule. We were off to a good start. As we concluded our interview, Frank made this touching admission after talking at length about his wife Susan's Alzheimer's: "As you can imagine, I live a pretty lonely life, Basil, and an intellectual conversation is a welcomed diversion, and I really enjoyed it."

Frank then called Jim Lovell, the third crew member on Apollo 8, who also gave me a revelatory interview. Like Bill and Frank, Jim, who is ninety-one, has the same luminous eyes, which still radiate with the readiness of the majestic symbol that could be found on the mission patches popular with the lunar astronauts: the eagle. With his golden retriever, Toby, sitting at his feet, Jim cited optimism as the most important quality a person can have and says it's what got him through the Apollo 13 crisis ("Houston, we've had a problem"), which was turned into a blockbuster movie. But Jim's most tantalizing reflection during our conversation pertained to birth and death. It is a singularly original thought, a contemplation that the dean of Yale Divinity School says he's never heard before and could only have resulted from Jim's two trips orbiting the moon.

Lovell's powerful formulation? "We don't go to heaven when we die, we go to heaven when we're *born*." There are other gems like this from the Eagles, as we'll discover later in the book.

Sensing that their final resting place in history is just around the corner, the Eagles opened their homes and their hearts in ways that I don't think would have been possible from their younger selves ten or twenty years ago.

They are intrepid souls whose wisdom has been forged by a voyage and a view of Earth that only they have experienced. "I grew up watching these guys," says Bezos, who told me in a heartfelt interview that he was directly inspired by the moon landings and believes that the Eagles' "elder statesmen" status and the "bluntness of age," as he put it, make their final reflections worth listening to.

"They have a very low tolerance for bullshit. And they're often very courageous in calling it out," says former NASA flight surgeon and psychiatrist Dr. Patricia Santy, who has seen (and studied) the psychological profiles of the Eagles.

As an example, some of them call out senators and congressmen who wear the American flag on their lapels, as "false patriots and self-serving jackasses." They remind us that courage, quiet patriotism, and conquering fear—the real *right stuff*—all emanate from deeper sources: a commitment to the common good, and belief in something greater than oneself.

They cry out for us to take care of the planet. They urge us to reframe our view of Earth to theirs: no identifiable nations, borders, or races. Just humans, planetary brothers and sisters riding on the Earth together.

Their wish for all of us is to keep pushing the boundaries (as they did) and always, always live life with fierce optimism and faith that, like the moon shot, any goal—no matter the odds—is as achievable as your resolve to see it through.

The

MISSION

OF A

LIFETIME

The Eagle preparing to dock with *Columbia*, July 21, 1969

Introduction

When Neil Armstrong and Buzz Aldrin took their first steps on the moon on July 21, 1969, their colleague Mike Collins, orbiting above them in the command module, worried about one thing: Would the ascent engine on their lunar module, the Eagle, work properly to get them off the lunar surface? "I was a lot more worried about getting them up off the moon," said Collins, "than I was about getting them down onto the moon. The motor on the lunar module was one motor, and if something went wrong with it, you know, they were dead men, there was no other way for them to leave."

Although few people knew it, President Richard M. Nixon had the same fear and was quietly dreading the other televised speech he might have to give. When Nixon placed his congratulatory phone call to Neil and Buzz during their two-hour walk on the Sea of Tranquility, he concluded his conversation with them praying "that [they would] return safely to Earth." There was much more behind the president's hopeful request for God's assistance in keeping the two Eagles safe. What Nixon couldn't let go of was the warning Frank Borman had given his speechwriter, William Safire. It was Borman who first reminded Safire of the possibility of "mishaps" during the first lunar landing and told him that he needed to prepare a contingency speech for Nixon. A contingency speech for what? Safire wondered. Safire, who would later become a celebrated *New York Times* columnist, recounted his conversation with Borman in 1999 on NBC's *Meet the Press*. "I remember when Frank Borman, who was the astronaut who acted as the liaison with the White House, called me up and said, 'You're working on this moon shot, you'll want to consider an alternative posture for the president in the event of mishaps.'" At first, Safire said he couldn't figure out what Borman was driving at. "And I didn't get it," Safire continued, "until he added—and I can hear him now—'like what to do for the widows.'"

Then it sank in. Safire would have to prepare a eulogy, which, if ever delivered, would have turned Armstrong's immortal "one giant leap for mankind" into a Shakespearean tragedy that would haunt the world in perpetuity. Safire went to the Oval Office and handed the contingency speech to Nixon just in case the two Eagles were left to a slow public death. It read:

> Fate has ordained that the men who went to the moon to explore in peace will stay on the moon to rest in peace. These brave men, Neil Armstrong and Edwin Aldrin, know that there is no hope for their recovery. But they also know that there is hope for mankind in their sacrifice.
>
> These two men are laying down their lives in mankind's most noble goal: the search for truth and understanding. They will be mourned by their families and friends; they

will be mourned by their nation; they will be mourned by the people of the world; they will be mourned by a Mother Earth that dared send two of her sons into the unknown. In their exploration, they stirred the people of the world to feel as one; in their sacrifice, they bind more tightly the brotherhood of man.

In ancient days, men looked at stars and saw their heroes in the constellations. In modern times, we do much the same, but our heroes are epic men of flesh and blood. Others will follow, and surely find their way home. Man's search will not be denied. But these men were the first, and they will remain the foremost in our hearts. For every human being who looks up at the moon in the nights to come will know that there is some corner of another world that is forever mankind.

Nixon's undelivered memorial serves as a powerful reminder today of the range and complexity of the risks the twenty-four Eagles faced with each of their nine voyages to the moon. "Death was always right outside our window," said Alan Bean, who was the fourth Apollo astronaut to walk there. He also feared the ascent engine might not start, which his commander Pete Conrad suspected was worrying him. "Pete looks over at me and he says, 'You're awful quiet over there, Al.' I said yeah. He said, 'Are you wondering if this thing is going to start?' And I said, 'Yeah, probably.' He says, 'Me too, but if it doesn't, we'll be the first permanent monument to the space program on the moon.'"

"The unknowns were rampant," said Neil Armstrong, who gave himself only a fifty-fifty chance of making a successful landing. "The systems in this mode had only been tested on Earth and never in the real environment. There were just a thousand things to worry about in the final descent. It was hardest for the systems, and it was hardest for the crew. It was the thing I most worried about because it was so difficult."

What was it in the backgrounds of the Eagles that allowed them to risk their lives in space, where no one had ventured before?

How did they conquer fear?

Where did their leadership skills and humility come from?

What was it about their shared experience that made them all see Earth as the Garden of Eden waiting to be rediscovered?

That's what is explored here with the remaining moon voyagers, who reflect on fifty years of lunar hindsight, what it has taught them and can teach us. None of these men asked for fame. They don't see themselves as heroes. They are, like the mythic heroes of antiquity, flawed—some more than others, and they humbly admit it. But they remain the only humans to have seen Earth from the moon, and from that mystical perch their minds were rebooted with an altered view of happiness, and the value of time, and above all, a newfound esteem for our home planet.

"Everyone who went to the moon came back a changed person," said NASA flight controller John Aaron, who was one of the people celebrated in the Oscar-winning movie *Apollo 13* for his critical role in getting the crippled spacecraft and its depleted crew home safely. "It was an experience that somehow caused them to reframe, and you could tell they had some kind of emotional, or religious experience as a result of it."

While some of the Eagles talked privately of a great awakening after their flights, their biographies and chronicles of the space age focused on the ordeals of spaceflight, its risks and technical intricacies—not the transcendental impact of lunar exploration.

By nature, the men of Apollo, mostly trained as test pilots, were not introspective. In their world, where split-second decisions were required to avoid catastrophe, reflective thought could be deadly. Their job was to maintain an icy resolve, "the right stuff," as author Tom Wolfe described it in his legendary book and movie of the same name, which has become a catchphrase for audaciousness and equipoise.

"The Apollo astronauts were the cream of the crop," says Dr. Santy. "They were remarkable men in many, many ways; their skill sets were incredibly high."

As for *The Right Stuff*, the Eagles, for the first time, comment on both the book and the movie, which they say could have gone further in exploring the layered complexity of genuine courage.

And what about fear? Dr. Santy gives here a little-known example of it when Neil Armstrong (known as Mr. Cool Stone to his fellow Eagles) was running out of fuel while scanning the lunar landscape for a boulder-free area to land on: "When they were landing on the Moon, they had a heart rate monitor on Armstrong. And one of the things that was absolutely remarkable to the doctors was that his heart rate was 160. This is practically unheard of. His heart was beating so fast that's what we would say are the physiological signs of fear."

So how did Armstrong work around his fear? It's a technique described first by Winston Churchill's doctor in *The Anatomy of Courage*, a book in which he analyzed fear in soldiers and pilots during World Wars I and II. It's the same technique used by the Apollo astronauts and, as we'll discover, the same method used today by the Navy SEALs who trained to kill Osama bin Laden.

Side by side with their physical courage were also examples of moral courage. This is particularly true of Borman and Anders, the former in taking on President Richard M. Nixon over his overtly nationalistic public relations plan for the first moon landing, and the latter in a showdown with the nuclear power industry over safety standards when Anders was chairman of the Nuclear Regulatory Commission.

The psychological profiles of the Eagles revealed two virtues that Aristotle maintained are vital to a healthy society: devotion to something greater than oneself and the pursuit of what he called *the common good*, which Anders cited as one of the reasons he was willing to risk his life going to the moon.

It is reassuring to know that these men are still filled with boundless confidence in America's ability to always resurrect itself from setbacks and crises.

The Eagles' stories offer refreshing hope. As Neil Armstrong said shortly before he died:

Like that of Ulysses, each of our lives is a miniature Odyssey, going to new places, seeing new things, understanding new ideas, and each penetrating the biggest unknown of all: TOMORROW! For each of us it should be and can be an exciting voyage.

Chapter One

THE REAL RIGHT STUFF: SELECTING THE EAGLES

Apollo 11 mission patch

KEY LESSONS

- Be humble
- Be decisive
- Be brave, but not reckless
- Take calculated risks

- Give your children responsibility at an early age
- Believe in something greater than yourself

Courage involves a risk-reward decision with reward factors to include "Duty, Honor, Country, and the common good."

Bill Anders

The world you enjoy today was made by these people most of whom came from often economically challenged backgrounds but had a passion about doing something that was important, in this case the space program.   *Tom Brokaw*

W est Point and its motto really shaped me," says Frank Borman. "I went there an eighteen-year-old kid from a small desert town, and it was a four-year period that really molded my character and my beliefs."

No matter which service they were in, the Army, Navy, or Air Force, the Eagles were guided by *duty, honor, country*, in their unrelenting climb to reach the apex of aviation's pyramid in the 1960s: the Astronaut Corps. That was the place to be. President Eisenhower, who created NASA in 1958, believed test pilots would make the best candidates for spaceflight because they had already been picked from the military's elite aviators for the most dangerous job of all: testing supersonic aircraft right off the drawing board.

"Test pilot experience was critical," said Ed Mitchell, the sixth lunar walker. "It was a profession with an esprit de corps and a lot of danger and pioneering spirit. And when you were at supersonic speeds at high altitudes, learning to survive that, and bring your machine back down, that's the fundamental task. And the higher and faster you flew, the more exciting and dangerous it became."

NASA put out five requests for astronauts to join the space program between 1959 and 1966. Mike Collins, who was the command module pilot for Apollo 11, was selected in 1962 from the US Air Force Test Pilot School. "That group of astronauts was by far and away the best I had ever been associated with," said Collins. "There really weren't any weak sisters in the bunch. They were just an amazingly competent, hardworking, really good bunch of guys." (See the quick reference table at the end of this chapter for a list of the twenty-four Eagles and the flights they were on.)

By the time the Eagles arrived at NASA's doorstep for final testing and selection, they were already the proven best of their generation, and the competition was fierce. It was as if the NFL had found 110 fully formed Tom Bradys and then made them compete with each other for seven quarterback slots. Those were the final selection odds for America's first seven astronauts (the Mercury Seven) chosen in 1959 for Project Mercury.

"I thought I had the best job in the world from the day I entered flight training until I looked on TV one day, and Al Shepard goes up in a rocket. He's gone higher than I've ever gone, and faster than I've ever gone, and most important, he's made more noise doing it. He's even on TV doing it…how do I get that job," Alan Bean reflected wistfully.

"It certainly sounded very challenging," says Buzz Aldrin, "and something that if other people wanted to be a part of this (and this was a noble national effort), then I wanted to be part of it."

The candidates had to be self-confident but not arrogant, brave but not reckless, independent but trusting of authority. As a result, no group of prequalified alpha males, with extremely high IQs, had ever been more tested, poked, prodded, interviewed, and psycho-analyzed than the Eagles. The foremost expert on the early psychological testing of the Eagles remains Dr. Santy, who in 1994 wrote an expansive, scholarly book, *Choosing the Right Stuff: The Psychological Selection of Astronauts and Cosmonauts.*

"What interested me most was the personalities of those extraordinary people who wanted to live and work in the dangerous environment of space or on distant planets," Dr. Santy writes. She worked at NASA's Johnson Space Center in Houston from 1984 to 1994 as a flight surgeon and flight doctor, where her suggestion to develop a more comprehensive set of psychological selection criteria for the shuttle astronauts got little support from NASA management. When she requested, as part of her research, the psychological data for the original Mercury, Gemini, and Apollo astronauts, the files were inexplicably "unfindable." It was no secret that NASA management and its astronauts had a historic aversion to psychological testing (hilariously recounted in Wolfe's *The Right Stuff*). At the beginning of the space program, however, it was seen as a necessary evil since no one knew how even the toughest test pilots would handle the confinement of a tiny space capsule surrounded by so many dangerous unknowns.

"I had little patience for the psychiatrists," recalls Borman. "We had to endure a battery of psychological tests, my first full encounter with the mysteries of psychiatry. There were the usual inkblot tests, which implanted in us the fear that if something looked like a tree, it was a definite indication that we were sexual deviants."

"They gave us inkblots to identify, while psychiatrists asked us crazy questions like, 'Who do you hate worse, your mom or your dad?'" wrote John Young, who walked on the moon's Descartes Highlands.

NASA, says Dr. Santy, did not want her to see those psycho-metric findings—not because there was anything negative to hide, but because they feared the risk of misinterpretation of the data, or possible leaks to the media. Dr. Santy, however, believed those files were necessary to help NASA refine new selection criteria to predict which personality types were best suited for long space voyages and colonization of distant planets like Mars. Dr. Santy can't say who, but she found an ally in her effort who was willing to give her the complete psychiatric evaluations and psychometric test results for the Mercury, Gemini, and Apollo applicants.

The files are a treasure trove. Except for Dr. Santy's mostly aca-demic audience, much of the information revealed here will be new to most readers. Because of privacy laws, the individual names of astronauts are never mentioned, and Dr. Santy's findings are gen-eralized to the group.

As she expected, the tests showed the Eagles were remarkable men on many levels and atypical of the male population in general, particularly when it came to rating their own self-confidence and overall optimism. The files show that 90 percent of those selected for the Mercury program came from loving, stable families and were firstborn or only children who'd never experienced any real failure in their lives. The other 10 percent selected had survived difficult childhoods through resilience and sheer willpower. "Some had stern disciplinarian fathers, and loving mothers, and the usual conflicts with siblings," Dr. Santy discovered. "What is not so typical is that very few of them had to contend with broken homes, and none had significant mental or physical trauma as a child."

NASA's key selection requirements for the Mercury Seven, says Dr. Santy, were that candidates "demonstrate good stress tolerance; an ability to make decisions; an effective ability to work with others; emotional maturity; and a strong motivation for *team* rather than personal objectives." The psychiatrist who pioneered the first psychological tests, made famous in *The Right Stuff*, was Dr. George Ruff, who in 1957 began working at the Aerospace Medical Laboratory as chief investigator of the stress and fatigue section at Wright-Patterson Air Force Base in Ohio. Ruff and his team had studied primarily military aviators and test pilots, the same group that President Dwight D. Eisenhower insisted the pool of astronaut candidates be chosen from.

Since no one had ever flown in space, Ruff and his team were dealing with a blank slate and zero scientific data for predicting the psychophysiological effects of seven-G launch forces, weightlessness, and other unforeseen sources of stress, which is why chimps were the first to make the trips into space, all with good results. To help NASA's Astronaut Selection Board find the best candidates,

Ruff, who died in 2017, developed the following eight psychological criteria, which really are timeless metrics for overall mental health and, at the time, were groundbreaking in their simplicity:

1. Candidates should have a high level of general intelligence, with abilities to interpret instruments, perceive mathematical relationships, and maintain spatial orientation.
2. There should be sufficient evidence of drive and creativity to ensure positive contributions to the project as a whole.
3. Relative freedom from conflict and anxiety is desirable. Exaggerated and stereotyped defenses should be avoided.
4. Candidates should not be overly dependent on others for the satisfaction of their needs. At the same time, they must be able to accept dependence on others when required for the success of the mission. They must be able to tolerate either close associations or extreme isolation.
5. The astronaut should be able to function when out of familiar surroundings and when usual patterns of behavior are impossible.
6. Candidates must show evidence of ability to respond predictably to foreseeable situations, without losing the capacity to adapt flexibly to circumstances that cannot be foreseen.
7. Motivation should depend primarily on interest in the mission rather than on exaggerated needs for personal accomplishment. Self-destructive wishes and attempts to

compensate for identity problems or feelings of inadequacy are undesirable.

8. There should be no evidence of excessive impulsivity. The astronaut must act when action is appropriate but refrain from action when inactivity is appropriate. He or she must be able to tolerate stress situations positively, without requiring motor activity to dissipate anxiety.

Not only did the Mercury Seven exceed the requirements of Ruff's criteria, but their IQ scores, which ranged from 130 to 141, were exceptionally higher than the national average of 98 (using the Wechsler Adult Intelligence Scale [WAIS]). The same high IQ scores would be true for the subsequent Eagles who joined the space program between 1962 and 1966.

What stands out in Dr. Ruff's reports is that "very few fit the popular concept of the daredevil test pilot." Daredevils had long been a feature in the American cultural landscape starting in the 1920s with wing walkers on biplanes doing stunts while flying under bridges; men (and women) inside barrels doing jumps over Niagara Falls; and other daredevil acts such as sitting on stacked chairs atop steel girders at the construction sites of the country's first skyscrapers.

Daredevils, some psychiatrists speculated, might have something to prove sexually, which in their minds was clearly not the *right stuff*. And, yes, in another surprise from the files, Ruff's team tested for sexuality in their interviews with the Mercury Seven:

Because of the possibility that extreme interest in high performance aircraft might be related to feelings of inadequacy
in sexual and other areas, particular emphasis was placed
on a review of each candidate's adolescence. Little information could be uncovered to justify the conclusion that
unconscious problems of this kind were either more or less
common than in other occupational groups.

In plain English, they were very secure males. By 1962, the
lengthy psychiatric interviews were stopped in favor of shorter ones,
and there would be no more questions ferreting out sexual motiva-
tions. Al Shepard, Gus Grissom, and John Glenn had successfully
proven that man could fly in space with no ill effects physiologically
or psychologically. They were national heroes on magazine covers
with an adoring public. At that point NASA decided it would be a
waste of time to collect extensive psychological data (pre- or postflight) that might wind up in the wrong hands and twisted to the
detriment of the space program—an irrational fear that Dr. Santy
says still exists at NASA.

What would continue for applicants to the program were the
final oral interviews, which Anders remembers with pure dread.
After weeks of physical stress tests and psychological probing, and
feelings of angst and uncertainty about the outcome, he had made
the final cut. This was his big moment! A chance to make his final
case, like a PhD candidate at an oral exam.

"I walked into this sterile room," Anders remembers. "There

was a row of guys sitting at a long foldout table, which included Al Shepard and Wally Schirra, along with several doctors and psychiatrists, who had my records stacked in front of them." Dr. Charles "Chuck" Berry, the head flight surgeon, opened the interview by telling Anders how pleased they were with his test results and that he was a strong candidate. But then Berry asked in a very concerned manner about a concussion Anders supposedly had five years earlier. "I never had a concussion," Anders thought to himself, "so I'm thinking, what's going on here?" Was this a trick question, Anders wondered, a red herring to see how he'd react, or a test of his honesty? "I couldn't bring myself to lie, so I thought, okay, I'll just say I've never had a problem with a concussion." He's since tried to find out from NASA, with no success, if they had confused his records with someone else's.

The other test Anders remembers vividly is the same one that flummoxed all the Eagles, an aptitude test on a machine like a massive video game with ten games played simultaneously for a score. Anders quickly figured out that the scoring system was based not only on getting the right answer, or the wrong answer, but on how long it took to make your choice. So the cumulative score was the number of wrong answers multiplied by the time it took to decide.

"I thought to myself, fine, if there's one of these ten games that's very difficult, then I'll just always hit A instead of killing time to decide if it's B, because zero times any number is still zero. I weeded out the three most difficult games and focused on the easy ones. I think my score was probably 150 percent better than anybody else's, mainly because the rest wasted time on the difficult games."

Pete Conrad, the third man to walk on the moon, known for his irreverent humor, enjoyed turning the tables on the psychiatrists, for whom he had little use. He came from a privileged, Main Line Philadelphia background and was a socially secure Princeton graduate who had no fear of jousting with the "shrinkers," as the Eagles referred to the psychiatrists. Conrad thought the Rorschach inkblot tests, which had been developed by Swiss psychologist Hermann Rorschach, were useless. The Eagles figured nothing good could come from free-associating about their "feelings" (almost a dirty word for aviators with the right stuff) by looking at a bunch of indecipherable designs supposedly configured to reveal your innermost thoughts and personality. To Conrad they were rubbish. When one of the shrinkers shoved a completely blank sheet of white paper in front of Conrad and asked him what came to mind, Conrad shoved it right back at him and said slyly, "I'm sorry, Doc. I can't."

"Oh....And why can't you?"

"Because you've got it upside down," Conrad responded. The joke was on the shrinker.

Charlie Duke remembers the inkblot tests with equal humor and dismissiveness: "And there was a joke at the time amongst all of us candidates to *not* say or see anything sexual *even if there's a naked woman in this inkblot...* 'don't see it, you know.' I see this, and I see that, *but no naked woman.*"

The lone holdout in this conspiratorial pact was, not surprisingly, Conrad, who couldn't contain his glee for another chance to rib the shrinks and turn himself into a Freudian's delight.

Everything was coming up genitalia. It didn't matter what inkblot card they put in front of Pete, he pretended to see nothing but female body parts and sexual acts. "Oh, a vagina. That's definitely a vagina," said Conrad, who handed the card back to the psychiatrist and waited for another one to deconstruct. "Hmmm. This is interesting.... This is a man and a woman doing it." Conrad paid for his irreverence by being rejected in his first application but was finally accepted in round two when the psychological tests were given less weight. The truth was he was an exceptional pilot with an exceptional mind.

In the end the tests showed that the astronauts who were finally selected all cited a love of adventure, "risk," and "pushing the boundaries," as they were fond of saying, but the differentiator from those not selected was their consideration of "calculated risk." Their interviews showed they all had an appreciation of danger, but with the following critical nuance according to Dr. Ruff's notes: They had "a conviction that accidents could be avoided by knowledge and caution. They believe that risks are minimized by thorough planning and conservatism."

"It's a very methodical, analytical type of process," Al Shepard, America's first man in space, said of flying experimental aircraft. "You know what they say: There are old pilots and there are bold pilots, but there are no old, bold pilots."

Much of this sophisticated risk analysis was the result of being highly educated. Of the twenty-four Eagles, all had bachelor's degrees (a NASA requirement), ten had master's degrees, two had

ScDs, one had a PhD, and five ultimately attended Harvard Business School's Advanced Management Program. Five went to West Point and six to the Naval Academy, both of which provided a free education. Not surprisingly seventeen of the twenty-four Eagles were Boy Scouts, eleven of whom the Scouts' website proudly points out were moon walkers. The Scouts' venerable motto, Be Prepared, was baked into NASA's DNA through the marvelous new invention of flight simulators, which the Eagles would spend thousands of hours practicing in for their missions.

Not having a college degree is what disqualified legendary test pilot General Chuck Yeager, one of the central heroes of *The Right Stuff*, from becoming an astronaut. Yeager, who was the first human to break the sound barrier and was played by Sam Shepard in the Oscar-winning movie version, trained a number of the Eagles as test pilots, among them Borman and Duke.

For Borman, who genuinely admired Yeager's flying skills, the real right stuff (not the movie version) required keeping your "ego" and "showboating" in check—and not letting it get in the way of the prescribed mission. Ego could kill you, or your aircraft. Borman says Yeager destroyed the NF-104A, which was a rocket-assisted Starfighter jet, because he let his ego cloud his judgment when he flew the plane beyond its known capabilities. The Eagles would never have done that, and Yeager, they thought, was bordering on being a daredevil.

"I thought he'd pulled a showboat stunt with an airplane designed for a specific purpose," writes Borman, "and setting altitude records wasn't on the agenda. What Chuck had done was wipe out the most cost-effective vehicle ever developed for preliminary space training." The Air Force canceled the NF-104A program as a result of Yeager's crash. For the Eagles the real right stuff required more mental discipline than that.

The Eagles were all children of the Great Depression, mostly born in the late 1920s and early '30s, a generation that journalist Tom Brokaw says began acquiring resiliency in their early teens:

> That generation of astronauts and the people who fought in the war were formed early in terms of what they could expect, and what they could expect is what they could earn, and that gave them a strong sense of can-do, of self-worth and an appreciation of the country in which they lived and the opportunities they were going to have. They came of age during the Great Depression when the expectations of what life would give you were very low, and whatever you got you had to earn, and it was hard intellectually, emotionally, and physically during that time.

It was not a generation that was coddled. They were given responsibility at a young age, and the Eagles credit that early empowerment with their ability to be decisive and not fold in the face of crises and setbacks. For example, Apollo 15's command module pilot, Al Worden, began driving the family tractor at age twelve to help manage the farm in Jackson, Michigan: "I came from a family of six kids, living on this small farm, and to make extra money I also worked during high school as a piano player in a band."

Jim Lovell, an only child, fell into the 10 percent category that had a rougher childhood than Worden and most of the Eagles. When he was in the fifth grade, his parents separated and his father died shortly afterward in a car accident. He took his first job at fifteen, baling hay for ten cents an hour to help his mom, who was making meager wages as a secretary in Milwaukee, Wisconsin. Lovell says that their one-room apartment was so small, the closet doubled as a kitchen and they had to share a bathroom with the other boarders on the floor. In his junior year at Solomon Juneau Business High School, he worked in the school cafeteria to further ease their struggles. "It's amazing how hard you would work in those days just to get three square meals," Lovell recalls.

"I lived my early years in Depression-hit Cartersville, Georgia, forty miles north of Atlanta, where my grandfather operated a filling station," said John Young, who was the ninth Eagle to walk on the moon. "I was hungry enough at age three to always eat an apple down to its core. With Mom going into the Chattahoochee mental hospital when I was five, I was growing up pretty fast by age seven learning to be responsible and get along well on my own, but with a lot of help from Grandfather Mank and Aunt Sarah."

Borman was fifteen years old, for example, when he swept floors every morning at Steinfeld's department store for $2.50 a week between six and seven o'clock. And then after school, he pumped gas and worked as a bag boy at Safeway grocery store. "They were measured at young age by what they could do," says Brokaw, "with the expectation that the job would get done, particularly those working on the family farm. It made them independent, inventive, resourceful and made them perfectly suited to become pilots and astronauts."

"I developed maturity and an appreciation of the work ethic by earning my own money," says Borman. "My folks never gave me an allowance. When I wanted or needed anything they considered justified or truly important to me, somehow they found a way. The paper route, however, gave me my first real feeling of the independence and pride that comes from contributing one's share."

Being given responsibility at a young age provided the Eagles with enormous self-confidence, but not arrogance, says Dr. Santy, which NASA was very careful to screen for. Of the thousands of outstanding pilots who applied for the Apollo program, those who made the final cut, she says, had to demonstrate meaningful self-awareness: "Sensitivity to self and others is kind of an ineffable quality that, actually, a lot of these guys possess. They're not aware of it, but it comes out in their leadership abilities and their ability to function in a group setting."

What's more, says Dr. Santy, "all of these guys believe in something greater than themselves, and that's a strong motivating force for them. They believe in *the common good*." And it is from this the Eagles say they drew much of their physical and moral courage. "Courage involves a risk-reward decision with reward factors to include '*duty, honor,* and *the common good*,'" says Anders, who believes that's what gave him the mettle to risk his life going to the moon for America, and before that defending the nation against the Soviet Union during the Cold War.

The notion of the "common good," the idea of believing in something greater than oneself, finds its roots in ancient Greece, says Harvard Classics professor Mark Schiefsky. "In Aristotle's *Politics*, for example, for democracy to work you had to some extent subsume your private interest to the common good and the notion that we're not just living for ourselves, but that we live for the community."

"From the time I entered the Navy," said Bean, who died in 2018, "I felt blessed by what this country had given me, and I felt it was my duty to serve America in whatever way I could. If it meant risking my life, I was okay with that since the space program was so important to America and to me. People need to think more about serving the *common good* instead of their own selfish needs."

Borman echoes Bean's sentiment with what he learned at West Point. "The academy doesn't teach arrogance; it teaches qualities like integrity, perseverance, and serving your country honorably, and those are timeless lessons for every generation of Americans."

For America's Founding Fathers, who were steeped in the classics, belief in the common good became a bedrock principle of American government. In *The Federalist Papers*, James Madison wrote: "The aim of every political constitution is, or ought to be, first to obtain for rulers men who possess most wisdom to discern, and most virtue to pursue, the common good of the society; and in the next place, to take the most effectual precautions for keeping them virtuous whilst they continue to hold their public trust."

These were not abstract principles for the Eagles, for whom the West Point motto kept popping up when they were asked to answer the most vexing question of all from the psychiatrists: Who am I? It was the one shrinker question the Eagles did not joke about. They knew exactly who they were. When asked to reflect on that question today, the Eagles have not changed their answer and remain steadfast in their commitment to selflessness and serving the greater good. Duke summed it up impeccably for all of them when he said:

> You take a stand on what you believe in and I think that comes from a sense of duty. If you have a sense of duty to your unit, or a sense of duty to the mission, then you will make a courageous decision. I was willing to accept the risk of death for the good of the country and the prestige of America.

THE TWELVE EAGLES WHO WALKED ON THE MOON

	Name	Born	Died	Mission	Alma Mater
1	Buzz Aldrin	1/20/1930		Apollo 11	West Point, Massachusetts Institute of Technology
2	Neil Armstrong	8/5/1930	8/25/2012 (Age 82)	Apollo 11	Purdue University, University of Southern California
3	Alan Bean	3/15/1932	5/26/2018 (Age 86)	Apollo 12	University of Texas at Austin
4	Eugene Cernan	3/14/1934	1/16/2017 (Age 82)	Apollo 10 and 17	Purdue University, Naval Postgraduate School
5	Pete Conrad	6/2/1930	7/8/1999 (Age 69)	Apollo 12	Princeton University
6	Charles Duke	10/3/1935		Apollo 16	US Naval Academy, Massachusetts Institute of Technology
7	James Irwin	3/17/1930	8/8/1991 (Age 61)	Apollo 15	US Naval Academy, University of Michigan
8	Edgar Mitchell	9/17/1930	2/4/2016 (Age 85)	Apollo 14	Carnegie Mellon University, Naval Postgraduate School, Massachusetts Institute of Technology
9	Harrison Schmitt	7/3/1935		Apollo 17	California Institute of Technology, University of Oslo, Harvard University
10	David Scott	6/6/1932		Apollo 15	University of Michigan, West Point, Massachusetts Institute of Technology
11	Alan Shepard	10/18/1923	7/21/1998 (Age 74)	Apollo 14	US Naval Academy, US Naval War College
12	John Young	9/24/1930	1/5/2018 (Age 87)	Apollo 16	Georgia Institute of Technology

THE TWELVE EAGLES WHO ORBITED THE MOON BUT DID NOT LAND

	Name	Born	Died	Mission	Alma Mater
1	Bill Anders	10/17/1933		Apollo 8	US Naval Academy, Air Force Institute of Technology, Harvard Advanced Management Program
2	Frank Borman	3/14/1928		Apollo 8	West Point, California Institute of Technology, Harvard Advanced Management Program
3	Mike Collins	10/31/1930		Apollo 11	West Point, Harvard Advanced Management Program
4	Ron Evans	11/10/1933	4/7/1990 (Age 56)	Apollo 17	University of Kansas, Naval Postgraduate School
5	Dick Gordon	10/5/1929	11/6/2017 (Age 88)	Apollo 12	University of Washington
6	Fred Haise	11/14/1933		Apollo 13	University of Oklahoma
7	Jim Lovell	3/25/1930		Apollo 8 and 13	US Naval Academy, Harvard Advanced Management Program
8	Ken Mattingly	3/17/1936		Apollo 16	Auburn University
9	Stu Roosa	8/16/1933	12/12/1994 (Age 61)	Apollo 14	University of Colorado, Harvard Advanced Management Program
10	Tom Stafford	9/17/1930		Apollo 10	US Naval Academy
11	Jack Swigert	8/30/1931	12/27/1982 (Age 51)	Apollo 13	University of Colorado, Rensselaer Polytechnic Institute, University of Hartford
12	Al Worden	2/7/1932		Apollo 15	West Point, University of Michigan

Chapter Two

THE TECHNIQUES FOR CONQUERING FEAR

Apollo simulators in which the Eagles trained for thousands of
hours to prepare for crisis situations

<hr>

KEY LESSONS

- Always stay calm
- Stay focused on what's in front of you
- Have backup plans
- Eliminate catastrophic, what-if thinking
- Never give up

Your life is on the line, and I thought I can't perform my job if I'm not calm!   *Alan Bean, Apollo 12*

Courage is basically understanding how to handle fear.

Frank Borman

To conquer fear is the beginning of wisdom.   *Bertrand Russell*

When Alan Bean looked out the window of his lunar module, the Intrepid, he saw nothing but boulders as he and his commander, Pete Conrad, approached the Ocean of Storms, their planned landing site on the moon. It was November 19, 1969. Bean, a former Navy aviator and test pilot, had trained six years for this moment, but unlike his two crewmates who had five spaceflights between them, this was his first time in space. And now Bean was scared. "I always thought of myself as one of the more fearful astronauts," said Bean, "and when I'd look out the window of the spacecraft, I would think, 'If that window blows out, I'm gonna die in about a second.' There's death right out there about an inch away."

A lot was riding on Apollo 12. It would be NASA's second lunar landing, and one of its primary goals was to see if Conrad and Bean could make a pinpoint landing. Neil Armstrong had been unable to do it and overshot his landing site by four miles, nearly causing him to abort the mission because the Eagle's computers and landing radar, it was discovered postflight, had software issues. MIT's engineers fixed the problem in time for Apollo 12, and NASA picked the Ocean of Storms to land on because it had a visually identifiable target to aim for: Surveyor 3, an unmanned probe sent there in 1967 to explore its surface using a robotic scooping arm rigged to a camera that sent back photos of its samples to Earth.

"I looked out," said Bean, "and there were so many more craters and rocks on the real moon than there were in the simulator we'd trained in, and I thought, 'There isn't a place to land,' and I got scared at that moment, and I felt my heart rate go up." It was exactly what confronted Armstrong four months earlier when his heart rate also spiked after spotting boulders cluttering his landing site at the Sea of Tranquility.

And then Bean and Armstrong did something to manage their fear that aviators have been doing instinctively since aerial combat began in World War I. No name existed for it then, or during the Apollo era, but today the Navy SEALs call it "front sight focus," says Dr. Eric Potterat, who served for ten years as the head psychologist for the US Navy SEALs, where he managed and implemented all the psychological and mental toughness programs for SEALs worldwide. When you engage front sight focus, you concentrate with singular intensity on the task, or mission, directly in front of you. Next, you immerse yourself *in the moment* to sideline your fear. And then, step-by-step, you begin methodically exploring every possible option to escape a deteriorating situation.

"It's all about controlling the human stress response; those that do it control performance in any environment," says Dr. Potterat.

"We all have negative thoughts, and how people navigate through those negative thoughts and arrest that process and reverse that process is extremely important. So when something bad happens, they put it away and stay mission-focused, and that was a paramount issue for the SEALs.... Whether you're an astronaut and you're doing a lunar landing—or whatever—it's all about mission, mission, mission! And you don't have time to worry about things that go wrong—you just have to maintain composure; you put it away in a black box in your mind."

Bean knew how to black box his fear. "And I thought, I can't perform my job if I'm not calm," said Bean, "so instead of looking out the window, I looked down into the cockpit and focused on the computer readouts, our fuel consumption, and calling out our rate of descent to the lunar surface while Pete fired Intrepid's thrusters to find a clear landing site." By focusing on the instrument panel and being in the moment, Bean said, he "managed to finally settle down." When he did look out his window, Bean finally saw Surveyor 3 a few hundred feet away. He and Conrad had nailed their pinpoint landing.

Armstrong's landing, by comparison, required incalculable grace under pressure right until the moment he touched down. In one ear he had the voice of CAPCOM (Capsule Communicator) Charlie Duke calling out his fuel consumption, which was running dangerously low. In his other ear was Buzz Aldrin calling out their rate of descent, along with two highly distracting computer program alarms sounding off (falsely) that they were in

Pete Conrad inspecting Surveyor 3, November 1969

trouble. Heightening the pressure on the thirty-eight-year-old Armstrong was the unavoidable presence of the world, listening live, to his moment-by-moment exchanges with Houston. Humanity was holding its collective breath during those final minutes, waiting to see if the two Americans would abort, crash, or have their names imprinted in history's ledger for eternity. With his heart rate climbing, and only eighteen seconds of fuel remaining, Armstrong compartmentalized the incoming data overloading his senses and *focused* instead on the only thing that mattered: steering the Eagle to a boulderless nest on the powdery surface below.

With all the technical jargon filling the scratchy-sounding broadcast, listeners at that moment weren't entirely sure what the outcome was until they heard Armstrong say, "Houston, Tranquility base here. The Eagle has landed." Duke shot back with a response revealing the extent of everyone's anxiety in Mission Control: "Roger, Tranquility. We copy you on the ground. You've got a bunch of guys about to turn blue. We're breathing again."

"Armstrong had all the physiological signs of fear except in his voice," says Dr. Santy. "You could hear him, and if you looked at the physiological data along with the voice data and he's saying, 'three feet over, two feet down,' his voice was completely calm."

Like Armstrong and Aldrin, Borman also followed the front sight focus script unflinchingly when the engine on the supersonic NF-104A Starfighter he was testing caught fire at 40,000 feet. He keyed into his instrument panel looking for solutions. He thought briefly about ejecting, but at a speed of Mach 2.2 (twice the speed of sound) it probably would have killed him: "In a situation like that you don't have time for fear. I focused in to see what could I do with this thing *right now*. The tailpipe temperature indicator showed a reading past the red line, and so I shut down the engine and started gliding as far as I could."

With his speed dropping to 250 knots, Borman realized he would never make it to the runway and risked restarting his engine to get him within landing range. Once again, the engine began overheating. But he was now close enough to the airstrip to shut it down a second time, allowing him to make a dead-stick landing to safety. "I never thought that there wasn't a way out, that there wasn't some way to salvage a situation," Borman concluded.

"I think staying focused is really important," says Charlie Duke, "and as a pilot, that's what you do. I can remember an instance when I was flying F86s and the main fuel control that regulated the fuel into the engine went out, so I had to go to the emergency one, and it was sensitive to throttle movement, and so I realized what was wrong and I was focused on flying it, watching the engine temperature and setting up this precautionary flameout landing. So, it wasn't fear that took over, it was a focused, logical progression of how to analyze problems and how to handle them."

"You deal with what's in front of you—especially in combat— because you never know what's going to happen next," echoes Aldrin, who flew sixty-six combat missions in the Korean War. He shot down two MiG-15s, was decorated with the Distinguished

Flying Cross, and got a doctorate in astronautics from MIT. "Combat gives you fear," says Aldrin.

In his analysis of courage in battle, Churchill's doctor, Lord John Moran, discovered certain common denominators that have remained timeless in their application for managing one's fear.

"The worst thing is seeing the flak, the flashes. You must leave your imagination behind you, or it will do you harm," Wing Commander Guy Gibson advised his crews before bombing runs over Germany.

By "imagination" Gibson meant "catastrophic, what-if thinking," which can lead to paralyzing fear and indecisiveness. If anyone had a right to catastrophize and visualize a slow death, it was Jim Lovell and his Apollo 13 crewmates, Fred Haise and Jack Swigert, when one of their oxygen tanks exploded shortly before they were ready to orbit the moon. Ever since the Apollo 13 movie, the phrase "Houston, we have a problem" has entered the American vernacular as shorthand for situations that are starting to unravel. What was actually said was slightly different and was uttered first by command module pilot Jack Swigert and then repeated within seconds by Lovell. The race to keep their ship from bleeding to death started fifty-six hours into the mission when ground controllers asked Swigert to stir the oxygen tanks, a routine procedure stirring the slush inside the tank, which would tend to stratify. When he flipped the switch to stir the tanks, faulty wiring in oxygen tank number two caused it to explode, rocking the Odyssey. Swigert in a calm voice (the trademark of all the Eagles) informed Mission

Control, "I believe we've had a problem here." The flight surgeon noticed from the transmissions on the biomedical packs taped to their chests that their heart rates were skyrocketing.

"This is Houston. Say again, please," came the reply from CAP-COM Jack Lousma. Apollo 13's mission transcript shows Lovell responding precisely seven seconds later. Within those seven seconds, Lovell says he thought that maybe they had been hit by a small meteor, as the Odyssey continued shaking and torquing from the explosion. The instrument panel lit up like a pinball machine gone mad with multiple caution and warning lights flashing everywhere. But it was not clear yet, either to Mission Control, or the crew, that the oxygen tank was the source of the blast. Lovell took a deep breath. With complete poise—and not a trace of panic in his voice—he said what may go down as one of the great understatements of all time: "Houston, we've had a problem."

John Young was in Mission Control when he heard Lovell's report. "And I thought when I saw that oxygen system leaking down, I figured we'd lost 'em, I really did. I didn't think we'd make it."

Reflecting now, Lovell says matter-of-factly: "The thought crossed our mind that we were in deep trouble, but we never dwelled on it. You instead recycle your mind and think about what you have to do *now*." With that, Lovell, Haise, and Swigert began focusing instantly on solutions to the cascading failures that were crippling their spacecraft.

"They're able to focus on the job at hand," says Dr. Santy, "despite the fact that physiologically they're reacting like any normal human being would do. They don't even think about, '*Oh I'm coping with my fear.*' They're just focusing on the task, doing what needs to be done."

"We never gave up or went into a fetal position and said what's going to happen if we don't get back, where are we going to be," says Lovell. "There's no high voices, no panic. People keep asking after all these years: 'Why were you so calm?' Because there was no alternative."

It was this very coolness, the lack of tension among the crew, that frustrated film director Ron Howard, who feared the three Eagles might seem too perfect or, worse, make for a boring movie. For dramatic effect he manufactured, therefore, a confrontation between Haise and Swigert in which Haise suggests, in a sneering exchange, that Swigert carelessly misread the instruments before stirring Odyssey's oxygen tanks, thus causing them to malfunction. "The crew conflict you saw in the movie wasn't there. I suppose they put it in there to spice it up," Haise told the *LA Times* in 1995. "We never said a curse word the whole flight."

TRAINING & CONTINGENCY PLANNING

Training and contingency planning factor heavily into mastering fear, says Dr. Potterat, who today serves as director of specialized performance programs for the Los Angeles Dodgers baseball team (he also holds workshops for corporate executives and entrepreneurs).

"When you talk about elite performers," says Potterat, "whether it's on a sports field, battlefield, in business, or outer space, it's about habitually training them to have multiple contingencies in place: Plan A, B, C, D, E, F, and this mitigates the stress response. So if Plan A goes wrong, you immediately revert back to Plan B. If that goes wrong, you have multiple contingencies in place, which keeps you mission-focused with no time to worry when things go wrong. This is what we call 'stress inoculation.'"

Current-day astronaut Chris Cassidy knows this contingency training well because he also happened to be a Navy SEAL (one of only three SEALs to become astronauts: William Shepherd and Jonny Kim are the others). "The training is very similar," says Cassidy, who was part of the SEAL team hunting for Osama bin Laden in Afghanistan's Tora Bora caves in 2001. "If you're flying into a compound with bad guys in it, you plan for every scenario and have alternate contingency plans backed up on each other to minimize the risks. So you put it all together and you feel confident that you can handle it." Regarding spaceflight, Cassidy says you're also simulating realistic failures. "I'm nothing compared to those guys," he humbly says, referring to the Eagles. But then, sounding very much like one, he adds, "You'll never leave the planet earth if you worry about everything."

"I believe that our training was so thorough," says Duke, "that we felt like we could handle any emergency. But we knew there were some emergencies you could not recover from. And you didn't let fear and worry overcome you because the probability was small, and so you focused on positive outcomes instead of the negative outcomes."

"NASA was very good at training for every contingency," says

Dr. Santy. "Training! Training! Training!" The Eagles say it helped build their confidence and manage fear.

"We were in the simulators every day," says eighty-four-year-old Harrison Schmitt, who received his PhD from MIT and was the only scientist-geologist to walk on the moon in what would be man's last voyage there in 1972. "You'd have a launch simulation and an EVA [spacewalk] sim and then you'd have a landing sim and a splash entry sim. You'd be going through that repetitively every day with the cycle intensifying in the last six months before your mission."

"I spent seventy hours a week, for three years, training for my flight all the way through," says Al Worden, who orbited the moon on Apollo 15. This intense schedule would be true for all the Eagles as they prepared for their flights.

"I think I had a little bit over two thousand hours in the simulator," says Duke, "where you learn how to handle these extreme emergencies, and so you get a mental checklist when something happens in the simulator, and you go through it over and over again for all the emergency procedures."

It was that kind of practice in the simulators, said Bean, that helped him manage his fear when he saw the boulder field out his window. "When I looked into the cockpit it looked like and felt exactly like the simulator on Earth, which I'd spent hundreds of hours in."

"Thorough planning, and thorough preparation, will lead to the best performance you can give," adds Duke. "I'm not saying you won't have a failure, because machines break and sometimes you can't recover. Airplanes explode, spacecraft crash, rockets explode. But if you've gone through your planning for your mission moment-by-moment, minute-by-minute and have a thorough plan—and then you train to that plan—and stay focused on that plan, then you're going to have the best chance of coming out with a good performance."

An entire group of NASA engineers, the SimSups (simulation supervisors), was dedicated to simulating every possible failure and putting the astronauts (and flight controllers) through carefully scripted daisy chains of survivable malfunctions, recalls SimSup Bob Holkan. Holkan was born on the Sioux reservation in Rosebud, South Dakota, and went to NASA shortly after graduating with a double major in math and chemistry from Southwestern State College in Oklahoma. "The way our scripts were put together was to always walk on the edge, but never to kill the crew. Don't put in a malfunction that you can't do anything about, because that's a waste of time. When you put those guidelines in, you're trying to train people. You're trying to make them confident. You're trying to make them competent. So, you're walking a fine line."

NASA used fifteen simulators to train the Eagles, and it was in the command and lunar module simulators that the crews spent 80 percent of their training time, logging in a staggering 29,967 hours according to NASA's history office. Each simulator consisted of an instructor's station, crew station, computer complex, and projectors to simulate stages of a flight.

Different projectors duplicated the scenery of space outside the spacecraft's windows. Each window of the simulator included a seventy-one-centimeter fiber-plastic celestial sphere embedded with 966 ball bearings of various sizes to represent the stars, the Earth, and the moon and its surface.

The flight controllers who manned Mission Control in Houston—and were largely invisible to the public—were, in fact, more critical to a mission's success than the astronauts in space. From their flight consoles they controlled every spacecraft system and computer program associated with getting the spacecraft off the ground, into space, on the moon, and back home safely.

Mission Control during Apollo 11 moon landing, 1969

They ran the show. If systems were failing, as happened on Apollo 11, 12, and 13, they had to find solutions within minutes of their occurrence. It was for this reason, says Holkan, that he and all the SimSups went out of their way to train fear and panic out of the controllers as well. They even created simulations that would unmask certain personality traits that might interfere with the outcome of a mission. "With an introvert," for example, says Holkan, "we'd script malfunctions into the simulation designed to see if he'd come out of his shell and speak up, otherwise the team and the mission would fail."

The story of the flight controllers is no less exceptional than the Eagles'. They were mostly "country boys" with engineering degrees from state universities, who were recruited by NASA right out of college. Computer science was in its infancy in the 1950s and '60s, and NASA was willing to roll the dice on young graduates eager to figure out how to fly to the moon. Everyone was on a giant learning curve. They were inventing, on the fly, engineering and management systems never seen before, with moving parts spanning multiple industries and disciplines across the country and around the world.

DOING YOUR HOMEWORK & DECISIVENESS

The fear-tackling methods of doing your homework and decisiveness reached their apotheosis when two flight controllers made two rapid-fire decisions that will be studied for as long as there is a space program and simulators remain a key part of the training. Both decisions were made within moments of the problem arising. Had the two controllers reacted differently, President Kennedy's dead-line of landing men on the moon before the end of 1969 might never have happened on time. •

Eleven days prior to Apollo 11's launch, SimSup Dick Koos told his technicians to load sim script number twenty-six into the simulators to test the flight control team who would be manning the consoles the day of the historic moon landing. It was the most prescient simulation Koos could have scripted. Three minutes into the simulated landing sequence, Koos decided to throw a curve-ball at Steve Bales, the lunar module computer system expert. Without any warning, Koos overwhelmed Bales and his team with multiple onboard-computer alarms to see if they would abort the landing. Bales, unable to sort through and understand the different alarms, decided to abort. In the debriefing after the simulation, Koos explained to Bales how he erred and why the 1201 and 1202 computer program alarms could be ignored in a live landing.

Nine days later, as Armstrong and Aldrin approached Tranquil-ity Base, program alarms 1201 and 1202 sounded off portentously, just as they did in Koos's simulation. Bales immediately began replaying the simulation in his head. He checked with his back-room team on his headset, and then, without hesitation, famously called out to flight director Gene Kranz (of *Failure Is Not an Option* fame), "We're go on that alarm."

President Nixon would later honor Bales with a NASA Group Achievement Award and praise him for his gutsiness. Introducing Bales, Nixon said, "This is the young man, when the computers seemed to be confused and when he could have said 'Stop,' or when he could have said 'Wait,' said, 'Go.'" Bales, son of a school janitor and a beautician, had achieved his own moon shot within four years of graduating from Iowa State University. He was twenty-six years old.

For John Aaron it wasn't simulator training that produced his nerves-of-steel moment that saved Apollo 12 (Bean and Conrad's flight) from aborting shortly after launch. It was that other fear-dampening prescription—exhaustive research—that gave him the courage to make the right call. On launch day, November 14, 1969, a storm front was moving in more rapidly than anticipated. President Nixon and everyone watching from the VIP stands reached for their umbrellas as the rain quickly got worse. Thirty-six seconds after the Saturn V rocket cleared the tower, Conrad radioed Mission Control with an ominous report.

"Okay, we just lost the platform, gang. I don't know what happened here; we had everything in the world drop out." Conrad's dashboard ignited with a blaze of warning lights telling him the main power in his command module was suddenly gone. The spacecraft's electrical systems were Aaron's specialty. As the flight EECOM (electrical, environmental, and consumables manager), it fell to Aaron to figure out what just happened. What happened, but no one knew it yet, was that Apollo 12 had been hit by not one but two lightning strikes. The Saturn V rocket turned into a gigantic lightning rod, causing Aaron's console (and everyone else's) in Mission Control to erupt into an unrecognizable pattern of numbers.

Aaron was an Oklahoma farm boy who, at age twelve, impressed his father with his talent for taking apart and rebuilding tractors. But Aaron was blessed with something far greater. He had a photographic memory harnessed to a passion for poring over and absorbing the schematics of the Apollo spacecraft. Aaron stared at his console. And he kept staring at it until he realized that he'd seen this pattern of numbers, dancing through his head, once before. Aaron was an untiring whiz kid who had a habit of showing up at Mission Control on his days off because of his unquenchable need to keep learning. A year earlier he was there witnessing a test in which an electrical system failure displayed a similar pattern of numbers that made no sense to him. Aaron was determined to figure out the pattern even though it wasn't his job.

"I had to understand why that pattern showed up," he recalls, "and so I spent the morning with an electronic engineer that I had a lot of respect for by the name of Dick Brown, and we figured out what happened. We figured out why that random set of numbers came up."

Armed with the homework he'd done, Aaron knew exactly which switch in the command module the crew had to throw to get its instrumentation back online. When he called out the obscure switch, SCE to aux (signal conditioning equipment to auxiliary), Conrad had never heard of it. "What the hell is SCE? And *where* the hell is it?" said Conrad, but luckily Bean knew exactly what, and where, the switch was, and he reached for it. Apollo 12 was on its way to the moon. The controllers and Eagles were so impressed by Aaron's quick, decisive call, from that moment on they referred to him as a "steely-eyed missile man." He was only twenty-four years old.

"Courage," Borman says, "is basically understanding how to handle fear." At the height of the Great Depression, when the Eagles were infants, President Franklin Delano Roosevelt told a scared but hopeful nation in his 1933 inaugural address that "the only thing we have to fear is fear itself." The Eagles understood this. The Eagles knew fear. They felt it, but they refused to be sidelined by it. NASA helped them manage that fear by creating a vast machine, manned by thousands of engineers, who simulated every conceivable scenario that might lead to failure.

"We had been trained to think don't ever give up as long as you've got options, and we never ran out of options," said flight director Gerry Griffin. Having multiple options was an article of faith at NASA, and they engineered those options into redundant systems that would kick in should any one part of the intricate machines they built fail. It helped make the fear go away. And then everyone practiced. And they never stopped practicing until they achieved their goal of landing on the moon.

Chapter Three

THE WIVES: BRAVER THAN THE EAGLES

Valerie Anders (*Courtesy of Valerie Anders*)

Susan Borman, age 20 (*Courtesy of Susan Borman*)

Dotty Duke holding lunar module (*Courtesy of Dotty Duke*)

Marilyn Lovell wears Jim's Christmas Day present, 1968 (*Courtesy of Lovell family*)

I think the wives had more courage. And I'll tell you why. The wives were perfectly aware that their husbands could die. And that was not something that their husbands ever thought about.

Dr. Patricia Santy

I have a little bit more respect for the SEAL wives than I do the SEALs. It's because of the ambiguity of not knowing about their missions, which made it even worse for them.

Dr. Eric Potterat

NASA's SimSups wrote brilliant simulation scripts to protect the Eagles from every imaginable crisis except one: failing marriages. The thousands of hours the Eagles spent in the simulators were thousands of hours not spent with their wives and children.

"They really didn't test the wives," says Valerie Anders, looking back on that period. "There was no one that said, 'How is your marriage?' Or, 'How does your wife respond to a crisis?' They just cared about the men." Many of the wives would pay with their sanity (and broken homes) for NASA's neglect.

"One day I did the math," says her husband, Bill, "and I calculated that I saw each of our kids ten minutes on the average, by themselves, per week." Bill still feels the guilt. "As a way to make up for all the time I didn't spend at their games and school events, I later bought each of them their own houses and helped pay for their kids' college educations."

The divorce statistic for the twenty-four Eagles was grim, with nearly 60 percent of the marriages ending in divorce. The wives suffered debilitating stress raising children with no help and facing an invasive press corps, all the while fearful for the lives of their husbands, who expected them to project a happy face to a watchful nation while they trained twelve to fourteen hours a day in service of their true passion—the moon. Their marital misfortunes were chronicled in *The Astronaut Wives Club*, a book that spawned a film and television series of the same name. Even today, fifty years after the Apollo program ended, the wives gather for reunions organized primarily by Valerie Anders. Their enduring bond was forged during their monthly coffee meetings to support each other in matters large and small, a sisterhood that would gather in each other's homes when it was their husband's turn to blast off on their perilous odysseys to the moon.

What has not been written about are the four remarkable marriages (with both spouses still living) that survived the Apollo era, and why they succeeded. The love stories that the Anderses, Bormans, Dukes, and Lovells reveal here are as stirring as they are varied. Each couple's path to marital success was special in its own way, and how they got there is a harrowing tale. It's the female version of the right stuff in the dawning era of 1960s feminism and the sexual revolution, which changed American mores and challenged the institution of marriage itself.

Not long after Frank Borman retired from NASA in 1970 and joined Eastern Airlines, his wife, Susan, checked into the Institute of Living in Hartford, Connecticut, for a four-month-long stay to recover from a mental breakdown and alcohol abuse. Every weekend, Frank says, he would fly to Hartford so he could be with her, taking long walks, communicating his love for her, and crucially, he says, renewing their marriage vows. Frank, never one to shift blame, fully admits that he was too career focused to notice Susan's alcohol addiction and downward slide into depression. But at an annual convention of NASA flight surgeons, where Frank was invited to be the keynote speaker, he went off script, and in his own cry of anguish—as Susan was still in recovery—he let the doctors have it.

"What have you doctors ever done for the wives of the astronauts? I'll tell you what you've done, absolutely nothing. This doesn't go just for my wife and my family but for every military wife and child whose problems and feelings you've been ignoring since time immemorial."

Other than assigning a protocol officer to the families when the astronauts' missions were under way, NASA had no playbook for the wives. No support groups. No counselor, or anyone other than a minister and other astronauts' wives, to comfort them if their husbands died in plane crashes or, in the most gruesome case, were incinerated as Ed White, Gus Grissom, and Roger Chaffee were when their Apollo 1 spacecraft caught fire during a launchpad test on the ground in 1967.

The Apollo 1 fire was a tipping point that introduced an entirely new dimension of fear for all the Eagles' wives. They were all intimately familiar with the odds of dying in the high-performance aircraft that their husbands were testing before they got to NASA. They were all used to the grim reaper shrouded in the infamous black car delivering the news of death. The sight of it slithering through military base housing meant one thing: a little boy or girl would never see his or her father again.

"When we were stationed at Hamilton air force base in San Francisco, six pilots were killed," Valerie remembers. "So, the *black car* would come through the neighborhood with a priest, or the minister, and the commanding officer and stop at a house. And, so, I kind of had that dread that that *black car* might stop at my house."

Neil Armstrong told his biographer that "in the year 1952 alone, sixty-two pilots died [at Edwards Air Force Base in California] in the span of thirty-six weeks, an astonishing rate of nearly two pilots per week, many of them involving flight test."

Valerie and the other wives understood the military and test pilot life. What they were not used to and had no real handle on—nor did anyone, for that matter—was the brave new world of spaceflight. Those anomalous space capsules (later called spacecraft) appeared claustrophobically lethal, like glittering high-tech caskets, which in one case they would become. The one-man Mercury capsule was barely larger than an old telephone booth (six by six feet), the two-man Gemini spacecraft the size of a Volkswagen Beetle, and the three-man Apollo vehicle about the size of a large, modern SUV.

Test pilots like Yeager joked that their astronaut occupants were nothing more than "Spam in a can" and that monkeys had made the first test flights; this last part was true, as "Ham the Chimp" demonstrated ably on January 31, 1961, that he could survive a sixteen-minute suborbital flight in space. When the Navy recovered his capsule and brought him on board the recovery ship, Ham was all smiles. He joyfully ate an apple and was happy to get back to his earthly routine. Until, that is, the very next day. Ham, an instant celebrity, was all at once face-to-face with a horde of reporters coming at him with a fusillade of light bulbs exploding right in his usually cheerful mug. Ham lost it! Fame was not for him. He hollered, beat his chest, and bared his teeth at the press corps—a sensation the first ladies of space would come to understand as they, too, were hounded by print and TV reporters for the eight long years they and their husbands were in fame's crosshairs.

Ham's capsule had no wings, no sleek fuselage, no ejection seats, no glide path to safety if something conked out as it did for Borman when he landed his crippled F-104 Starfighter. Worse, those cylindrical cones were nestled like fragile eggs on top of retrofitted ballistic missiles meant for war, which in the pre-manned, pre-Ham launch tests had an unnerving tendency to blow up or, in some cases, fizzle like a wet firecracker right on the launchpad—"Kaputnik" as *Time* magazine dubbed one failed Vanguard launch on December 6, 1957.

And yet, despite the early disasters, the wives recognized that their husbands were riding a wave of technological innovation that was harnessed to the greatest engineering effort in the annals of human exploration. Things could only improve, they reasoned, if four hundred thousand engineers and the best PhDs on the planet were working to make the Mercury, Gemini, and Apollo spacecraft as death-proof as possible. Adding to their comfort was that everything was taking place on live television. Surely, they thought, NASA wouldn't risk all that publicity and potential for televised catastrophes if those rockets and spacecraft wouldn't hold together better than some of the experimental jets that had a habit of digging their husbands' graves into the Mojave Desert floor surrounding Edwards Air Force Base. On top of that they were becoming famous too—America's first iteration of reality media stars. Their proud smiles exalted *Life* magazine's covers as icons of American femininity and motherhood supporting their right-stuff husbands, who were laying their lives on the line to reach the moon before the Soviets.

It is easy to overlook the importance of President Kennedy's decision to make America's space program totally transparent. Unlike the obsessive secrecy of the Soviet space effort, Kennedy told the nation, "We take an additional risk by making it in full view of the world, but as shown by the feat of astronaut Shepard, this very risk enhances our stature when we are successful."

And so it did. It was a brilliant piece of public relations and global diplomacy on Kennedy's part that showed America's exceptionalism, because as Kennedy would go on to say: "We go into space because whatever mankind must undertake, free men must fully share."

Dotty Duke says it was that global spotlight, backed by NASA's mighty engineering machine, that eased some of her fear: "We really trusted NASA. As far as him dying, Charlie, like all of them, didn't think about death and we didn't talk about death. I never even really thought about it and I told people, how could I be worried about a husband that has four hundred thousand people working to make sure he would have a successful flight instead of say, being in Vietnam, and getting shot at, so, I was not one that was really very concerned."

Dotty's optimism was grounded in reality. From 1961—starting with Al Shepard's first flight—to 1967, NASA had a total of sixteen successful manned launches—all of them televised and followed by parades, cover stories, and visits to the White House. Yes, there were a few close calls during some of the flights, but no one blew up...except, paradoxically, in the old-fashioned way. There were five astronaut deaths during that time, but four were the result of crashes in NASA planes, and one from a car crash. None had anything to do with the spacecraft or training for the missions.

And then reality smacked everyone in the face with a primal

scream for help that not only broke NASA's perfect track record but also ultimately shattered the already fragile confidence of many of the wives, eventually leading to one suicide and for many a downward spiral into Valium and alcohol abuse to ease their very justifiable terror. At 6:31 p.m. (EDT) on January 27, 1967, the newest of NASA's spacecraft, Apollo 1, the technological marvel that would take man to the moon, became a crematorium during a routine test on the ground.

"And all at once I heard it, *fire*! We heard the shouts from the crew," said John Aaron, who was listening to the test in the control room.

"Because I had been in the flight test business quite a while I've seen death happen in various ways, but not like that," recalled Chris Kraft Jr., NASA's first flight director. It turned out that a stray spark from damaged wires ignited a fire in the command module, which was filled with pure oxygen. Within seconds, the temperature in the capsule reached 2,500 degrees Fahrenheit (1,371 degrees Celsius), and the interior cabin pressure doubled in twelve seconds. The crew never had a chance.

Ed White's thirty-four-year-old wife, Pat, was close friends with Susan Borman, who reported that for weeks after the funeral, Pat kept saying to her, "Who am I, Susan? Who am I? I've lost everything. It's all gone." When Pat tried overdosing, Susan and the other wives rallied to stabilize her and filled in to help with her children and life's daily demands. (Pat would later commit suicide in 1991.) Ed had been the first American to walk in space, in 1965, and was described by Frank Borman as "the astronaut's astronaut, a handsome and powerfully built man who actually seemed indestructible." He continues: "He and I were very much alike in our devotion to our families. When Ed died, it hit Susan—for the first time with that much strength—that I wasn't immortal either."

THE ANDERSES

Bill and Valerie near their home in Anacortes, Washington *(Courtesy of Bill and Valerie Anders)*

Bill and Valerie are in their sixty-fourth year of marriage. It's a milestone few people reach, and both of them continue to be graced with good health. She is eighty-one and Bill is three years older. Valerie says Bill practiced two things during the Apollo years that she considers essential in marriage: communication and inclusiveness. From the beginning Bill decided to make Valerie a full partner in everything he was doing at NASA—even taking her with him on factory tours so she could see, from the ground up, the assembly of the Apollo spacecraft that would take him to the moon.

"He was pretty candid about the risk, but he was also very good about sharing how things were going, mechanically and otherwise. I went to tour where the spacecraft was being built. I went on a lot of tours with him. So, he was very open about what was happening and how he felt about it. He knew there was a risk and I knew there was a risk, but as I said, he worked so hard for that opportunity that I wasn't about to say, 'No don't go,' because I'm worried."

"As I look back," Bill says, "one of my lucky decisions in life was marrying Valerie. We met very young. She was nineteen when we married and she became the arch housewife, and as I got consumed by my job initially as an interceptor pilot, followed by Apollo, and then later in business, she just became both father and mother and did everything."

The Anderses' love story began as it did for so many of the Eagles, in their teens. Their childhood experiences were, in fact, worlds apart. Bill was born in Hong Kong in 1933, which at the time was under British rule. His father, Lieutenant Arthur Anders, was a naval officer, and his mother, Muriel, was a member of the Daughters of the American Revolution. Lieutenant Anders was transferred to the USS *Panay*, a gunboat patrolling the Yangtze River in China, when the second Sino-Japanese War broke out in July 1937. As was the custom for Navy families, Bill and Muriel followed him from city to city in safe housing along the Yangtze as his patrols progressed.

The war nearly killed all of them. Young Bill was almost trampled underfoot when, as a four-year-old, he witnessed what remains his earliest and most searing memory, the rapid advance of the Japanese army into Nanjing in 1937: "My mother was dragging me through a massive crowd escaping the Japanese, where we were almost crushed and separated, but she kept holding me tight and we managed to get on a troop train heading south out of the city." Over a six-week period the Japanese would systematically rape as many as twenty thousand women and murder more than three hundred thousand Chinese in what historians now refer to as the Rape of Nanjing.

While Bill and his mother were escaping on the troop train heading to Canton, his father's gunboat came under attack by Japanese warplanes, nearly killing its captain and forcing Lieutenant Anders to take command of the ship. With his father fighting for survival, Bill and his mother, now in Canton, wound up being no safer there as the Japanese continued their attacks all along China's eastern coast.

"And then we got bombed in Canton, and the lights went out in our hotel room," Bill remembers. "My mom grabbed me again and we raced downstairs to the hotel's veranda figuring we'd be safer outside. And there was this big polo field right in front of the hotel and I could see the Japanese planes approaching on the other side of the Pearl River a couple of hundred yards away, and they started bombing and I saw all the explosions and fire."

By all rights that scene should have produced terror in a normal four-year-old, but not Bill. His mind's eye lit up in wonderment as he tracked the planes and the cinematic pyrotechnics surrounding him. "Looking back," he says, "I thought it was great, airplanes buzzing overhead, diving all around us, fire coming out, bombs going off. It was incredible." While holding his mother's hand on the veranda of the hotel in Canton, Bill had revealed the first traces of the right stuff. The little boy kept his cool and absorbed the action with the same defiant eye he would later show a Soviet bomber crew when he flipped them the bird for flying too close to his interceptor plane during routine patrol off the coast of Iceland.

Having survived the Japanese onslaught, Arthur was awarded a Purple Heart and the Navy Cross. The family was then sent back to the States, where Bill would eventually spend his high school years in San Diego, excelling in biology and geology. He won a place at the US Naval Academy, and it was after his first year there on summer leave, in 1952, that he met Valerie Hoard on a double date at the beach.

Unlike Bill's ringside seat to mayhem, Valerie grew up in what she describes as a little piece of paradise in the small town of La Mesa, California, ten miles east of San Diego: "It was a lovely childhood. My mother's parents, who were from Munich, lived next door and between their house and our house we had fig tree gardens and every kind of fruit tree you can think of, and chickens and rabbits. My father was a California highway patrolman. He rode a Harley-Davidson and I'd get on the back of his motorcycle and he'd ride down behind the house into the garage, and he'd turn on the siren for me and ride into the garage."

On their daylong double date, swimming and walking around Mission Bay, Bill and Valerie wound up being more attracted to each other than to the people they came with. Bill, it turns out, had a formal side to him, and knowing Valerie was special, he wanted to lay the groundwork, properly, with her dad. So nineteen-year-old Bill put on his midshipman's uniform and went to ask Valerie's father if he could start dating his sixteen-year-old daughter.

"And my father said, 'Go ahead, she's been going out all summer. Just have her home by...' I forget what the curfew was, twelve or something. So, we had two dates only," Valerie says, "because he was going back to the academy. On our first date we went to the Admiral Kidd Officers Club for dinner and then the Starlight Opera, which used to be really nice in San Diego. Next date was the officers club, and the Old Globe Theatre for a Shakespeare play, and I thought, what a cultured man." At the end of the dates, Bill, ever the gentleman, would shake Valerie's hand instead of kissing her.

Bill was also a prolific letter writer. When he returned to the Naval Academy, he knew he was in love, and would write Valerie daily letters ranging from his career hopes to thoughts on philosophy.

During the courtship, Muriel, however, was intent on having Bill marry an admiral's daughter. "His mother used to line up all the invitations from the admirals to have tea with their daughters," Valerie remembers, "and he would never do it, so his mother had a little grudge against me because I interfered with his Navy social life." A year later, when Valerie was seventeen, she and Bill got engaged.

During the military and NASA years, Valerie understandably kept worrying for Bill's safety. According to Dr. Santy, one of the great ironies for the wives was the following dichotomy: what kept their husbands safe on their missions—their ability to lead with their intellect as opposed to their emotions—proved disastrous in their marriages. "Take somebody like Neil Armstrong, for example, you couldn't find a less expressive man," says Dr. Santy. "He was more the *John Wayne*, strong silent type."

"That steadiness and that fortitude didn't carry over well when it came to being emotional," agrees Valerie, who remained close to Neil's first wife, Janet, until her death in 2018.

"Neil was the ultimate test pilot. You talk about how emotions had to be quelled and all that. He was steady as a rock. Just did these amazing things. But when he got home there was nothing emotional left. And they lost a daughter from a brain tumor when she was two. And Janet never recovered," says Valerie. "And Neil just said to her, 'Janet, you've got to get over it.' And I mean, she never got over it. And he couldn't cope with the emotional side of it, and eventually she divorced him."

Valerie was lucky because Bill had already proven himself as a communicator during his Annapolis days with his daily letters. He even made an audiotape for her to listen to (expressing his love for her and the children) should he fail to return from the moon. She was also fortunate to have the support system of her parents, who lived nearby, to help with the loneliness of not having Bill around much. Valerie says there were meditative retreats she took advantage of, but unlike the extensive science surrounding today's mindfulness movement, she did not find them helpful. Nor was she religious.

So what did Valerie consider the right stuff for being the wife of an astronaut? "General stability, I'd say, and just the strength that it takes to raise children…and not being resentful for doing it alone. I felt my responsibility was at home and to my children, and to do everything possible to help Bill fulfill his dream of going to the moon."

It was a marriage bargain that worked. Sacrifice and commitment on Valerie's part, inclusion and constant communication on Bill's part. Her life with Bill would take her on a privileged journey that included White House visits with both Presidents Johnson and Nixon; a year in Norway, where Bill served as the US ambassador; and ultimately to General Dynamics (GD), where he served as chairman and CEO. Bill considers GD (not Apollo 8) his greatest personal success. Many business case studies—including one from Harvard Business School— still cite GD's turnaround as one of the greats in business history… and today Warren Buffett, who served on GD's board, continues to praise him: "Bill Anders completely turned around General Dynamics. Every move he made was smart and owner-oriented."

Bill's success at GD, says Valerie, opened doors for her to help charities, including her service as a board member and donor to the Smithsonian Institution. Valerie and Bill continue to be each other's matching half, and along with their six children and thirteen grandchildren, they say they have much to be thankful for.

THE BORMANS

Susan and Frank making a donation to the Billings Clinic Foundation (*Courtesy of the* Billings Gazette)

As Susan Borman began succumbing to Alzheimer's, she wrote Frank a powerful note of gratitude for his love and the peace he brought her while she recovered at the Institute of Living. Frank keeps the note near his reading chair, a lifeline to the old Susan he describes as "blond, beautiful, and brainy." In the United States today, 5.7 million Americans of all ages have Alzheimer's. The disease devastates its victims by wiping out all functional memory and traumatizes their caretakers who watch as their loved ones are gradually reduced to feeble strangers.

While looking at Susan's note, Frank, in a rare moment of vulnerability, hung his head in despair. The commander of humanity's first mission to the moon, who always found a way out of every problem, admitted to feeling helpless for the first time in his life: "I was never in a situation where it was impossible to formulate a way out except for now when my wonderful wife, who's been in a nursing home for four years, and there's no way out with Alzheimer's. And that has been the most difficult thing of all for me to come to grips with because I always thought there was a fix for anything. And the best doctors that I can find tell me that there's no hope. It's far more daunting than anything I ever faced in the air force or NASA."

Frank and Susan started out strong and fast. They met in 1946 at Tucson High School, in Arizona, when she was a sophomore and he was a senior playing quarterback for the varsity football team. Susan was not your typical Depression-era high schooler. Her father was a surgeon and her mother a college graduate and dental hygienist. Susan was voted the most beautiful girl in the school, which was a little intimidating to Frank, who was too shy to ask her for a date. But he had a solution. In what was probably the only mischievous ruse of his life, Frank convinced a friend to call Susan (pretending to be him) so he wouldn't have to hear a potential rejection. He needn't have worried. Susan accepted and was drawn by more than his athleticism. Athletes were nothing exceptional. But solo aviators who were Frank's age? This guy, she knew, was different.

Frank's parents had it tougher. Frank was born in 1928, the only son of Edwin and Marjorie Borman, in Gary, Indiana. When Frank was five years old, his severe sinus problems forced the family to move to Arizona, where he could breathe better in the drier climate. His father's leased Mobil gas station failed in the teeth of the nation's financial tailspin, eventually forcing him to drive a laundry truck, while his mother took in boarders so they could hold on to their home. Frank never felt deprived. His dad, an automobile buff, was also enchanted by airplanes, and one day splurged on a five-dollar biplane ride for himself and Frank: "I sat next to Dad in the front seat, with the pilot in the cockpit behind us, and I was captivated by the feel of the wind and the sense of freedom that flight creates so magically."

Frank managed to get mostly straight As in school, which would carry him all the way to West Point. Susan, who was equally ambitious academically, went to the University of Pennsylvania to become a dental hygienist, like her mom. Frank says she had a maturity beyond her years and a down-to-earth quality unaffected by her beauty. She caught the attention of the Ford Modeling Agency, but Susan was not tempted and turned down their contract offer. While Susan embarked on her studies, Frank launched himself on the trajectory that would become his lifelong pattern:

putting work ahead of everything else. He became so focused on succeeding at West Point that he neglected his relationship with Susan and placed it on hold until he graduated. Susan felt rejected but remained hopeful. She sensed that the duty-bound West Pointer had a singular destiny, and was moved enough by his impassioned plea for forgiveness after his graduation that she accepted his marriage proposal. They exchanged their Episcopal vows in 1950 and went for a one-week honeymoon to the Grand Canyon, whose vistas Frank still finds more spectacular than any view of the moon from his Apollo 8 command module.

THE MILITARY LIFE & NASA

The hourglass on Susan's reservoir of resilience started its long, steady drain as soon as Frank reported to Nellis Air Force Base in Las Vegas for advanced fighter jet training in 1951. Nellis quickly gained the reputation for losing more pilots to training than the Air Force was losing in Korean combat. The United States military and its pilots, in those early years of jet fighter development, were prepared to take significant losses in the Cold War battle for better weaponry against the Soviet Union. In one week alone, Frank says, six of his fellow aviators were killed in accidents. Susan was eight months pregnant with their first son, Fred, when all the crashes were occurring. And they would keep occurring. Unmercifully. They were attending pilot funerals at least once a week. Each time the black car appeared at someone else's door, the grains of sand in Susan's hourglass got shallower and shallower. After the Apollo 1 fire, Frank says NASA's Dr. Berry began quietly dispensing Valium and sleeping pills to the wives in distress.

Susan chose alcohol instead. Frank blames himself for not recognizing sooner Susan's dependence on alcohol and the growing signs of her depression: "I never took into account the enormous sacrifice I was asking her to make, and I still feel guilty about that. I didn't reflect on how my sense of mission affected Susan and my sons. It wasn't that I loved them any less or wouldn't do anything in the world for them, but the focus on what I was trying to do for NASA, the Air Force, and at Eastern Airlines, always took first place."

Susan was practiced at hiding her loneliness, fear, and frustration from Frank. She knew, as all the Eagles' wives knew, that it was their solemn duty to not burden their husbands with family problems or complaints that might cause them to lose their focus while flying their missions. It could kill them. It said so right in *The Air Force Wife*, a 362-page book that was required reading for all the wives:

> It is said that domestic troubles have killed more aviators than motor failures, high tension wires and low ceilings, so as an Air Force wife your responsibility is great and your job is of big proportions if you live up to the finest traditions of the service.

This was one part of the handbook the wives took seriously, and while they were able to convince their husbands that they were okay, they weren't able to fool each other. Dotty was one who suspected that Susan was getting shakier. "When I look back, the way she would dress and the way she would treat her hair, or just take care of herself, you could see she was fragile and going through depressions back then."

By the time Frank was ready to make history on Apollo 8 in 1968, Susan made the mistake of asking Flight Director Kraft how risky the first manned mission to the moon would be. And Kraft made the mistake of indulging his trademark need for dishing out the facts. Kraft had grown to respect Susan, but her grace, composure, and beauty lulled him into thinking she could take the answer. "Fifty-fifty" is what he told her...to which she kept a poker face and never confessed her terror to Frank.

Shortly after the launch, Susan sat down at the kitchen table and began writing out Frank's memorial service, convinced he would die. Her defenses had disintegrated to the point where she couldn't conceal what she was writing even from her fifteen-year-old son, Edwin. "He might not come back," Susan told him when he walked in on her. In a reversal of roles, Edwin became the calm parent, took the pen out of her hand, and told her, "Just remember, Mom...Dad gets to choose the way he goes—you and I don't have that privilege."

Other than that moment of weakness with Edwin, she never let on to anyone her nightmare scenario of Apollo 8's crew becoming a dead satellite of the moon. Years later, in a PBS interview, she talked about the burden of having to maintain a perfect public facade: "They had people looking into the background of the men, [and] they also had people looking into the background of the wives because they didn't want an oddball . . . it wasn't discussed, it wasn't written, but you had better be in every sense of the word, the All American Family in everything you say and do! We kept it like *Leave It to Beaver.*"

After twenty years of Susan watching her husband climb into supersonic cockpits and spacecraft, her deathwatch came to a long-awaited end when Frank announced his retirement from NASA following Apollo 8. Her fear was finally gone, but not the drinking or the masked depression, which Frank admits were still invisible to him. What Susan needed at that point was Frank—all to herself. No more weekend marriage only, catching snippets of his love and devotion. Instead she got more of the same as she found herself competing suddenly with Frank's newest mission, the presidency of Eastern Airlines, which was no less important to him than beating the Soviets to the moon. When her two sons left for West Point, and she was left to herself, the hourglass finally ran out. There was nothing left in her. Frank was faced with an inconsolable Susan who would not leave her bed. Her nerves had given out. With the advice and help of one of Eastern's doctors, he took her to the Institute of Living, where the senior psychiatrist, Dr. Richard Brown, not only saved Susan but also elicited from Frank his accountability in her illness. In multiple sessions with Dr. Brown, Frank came to see that his "mission versus family" mind-set was a false choice. "I should have spent more time with her and communicated better," says Frank, "and not been blind to the impact my devotion to mission and duty had on her. She needed me and I wasn't there."

Frank kept his promise to make Susan his new mission. After her recovery, Susan never touched another drink and neither did Frank. For a time she did drug prevention work in Miami, and Frank began including her in his work at Eastern, where he instituted the first employee alcohol-counseling program. After Eastern Airlines was sold, Frank and Susan moved to Las Cruces, New Mexico, to help their son Fred with his car dealership, while Edwin became a helicopter test pilot. Their retirement years in Billings, Montana, were blessed until Susan was diagnosed with Alzheimer's, eventually requiring round-the-clock care. "I hope to stay healthy," Frank says now, "so I can take care of Susan." To that end, Frank maintains his early morning exercise regimen and then goes to be by her side. During each visit, he tells her how much he loves her, even as she increasingly fails to recognize him. Every week he goes to church praying for a cure that may never come in time.

"She means everything to me, and I am convinced—and I don't think it's self-delusion—but I'm convinced that Susan and I will be together again forever. I have no idea—no concept of what the afterlife will be—but I have this faith that we will be there together."

THE DUKES

Dotty and Charlie at their home in New Braunfels, Texas, 2018
(Courtesy of Dotty and Charlie Duke)

As with the Bormans, "faith" is central to who Dotty and Charlie are today. God intervened during Dotty's bleakest moment and saved her life. One of the first questions that suicide prevention counselors ask to determine the seriousness of one's intent is: Do you have a plan? Dotty had one.

Unlike Susan, Dotty believed that Charlie would survive his flights, but very much like Susan, she felt like a widow because Charlie was absent both physically and emotionally. It was the cumulative weight of twelve years of endless solitary nights and needy children that eventually led to her suicidal depression: "The hardest part," Dotty recalls, "was his being gone so much, and they were gone more than they had to be sometimes. I felt abandoned with no attention or affection from Charlie, or much help raising the kids."

Even with all her anguish and heartache, Dotty was as support- ive as *The Air Force Wife* handbook told her it was her duty to be. She was proud that Charlie had been selected as an astronaut and even prouder that he would walk on the moon, but the handbook required her to put a lid on her vexations. "You want your husband to do well in his job, and so I looked at that as a wonderful oppor- tunity and I supported him the way I was supposed to support him. The military trains you like that. I had my duty to the kids. And I had my duty to my family. And I had my *duty* to NASA—and the press."

Since the space program was a new creature, NASA had no track record of advice for handling reporters, except to be positive and not lose it like Ham, which the wives often felt like doing. At one of their monthly coffee meetings, they decided to come up with their own media strategy. They created talking points good for almost all occasions. They reduced them to three crisp words hoisted on three separate placards for the press to photograph. Their bumper sticker simplicity had a touch of Madison Avenue to it. What they came up with was *Proud*, *Thrilled*, and *Happy*. Every time a mic was shoved in their faces and they were asked how they felt as their husbands were blasted off the launchpad in a ball of fire, the wives chanted their refrain . . . *proud*, *thrilled*, and *happy*. Proud? Definitely! But thrilled and happy? Mortified and distraught were more like it.

They were not living the *Leave It to Beaver* life. In all-American TV families, Mom and the kids never considered that Dad might not return from his nine-to-five desk job. Those dads were home for dinner every night solving quaint sibling squabbles like whose turn it was to mow the lawn or why homework wasn't done on time. Astronaut kids were living in a different universe. Dad wasn't on some commuter train crawling along at sixty miles per hour. He was orbiting the Earth at 17,550 miles per hour, a surreal five miles per second. On a gut level, the Eagles' kids understood that their dads were in a scary business. Dotty's six-year-old son, Charles III, for example, didn't think the rocket launches were cool at all. "The fire coming out of the rockets scared him," says Dotty, "and worse, he disliked being cornered by reporters all the time."

Eric Anders remembers well how the reporters would chase him too. "I have memories of running outside when the press was milling around our driveway right before the launch. A picture of me smiling and with my thumb up ended up on the front page of some major newspapers with a caption something like, 'Astronaut's Son Gives Apollo 8 the Thumbs-Up.' Even at four," says Eric, "I think I had a strong sense that my mother had to hold a lot in. Later I would learn that that was standard operations for fighter pilots' wives at the time—especially given how so many pilots of supersonic jets were dying due to how dangerous those planes were then."

"I was in second grade. I remember that we did not go to the Cape to watch the launch," Eric's sister Gayle said. "I found out later it was because, #1, they could not afford it (money was always tight with five kids and a military salary) and, #2, my dad did not want us all out in the grandstands in case the rocket exploded during the launch. How could my mother comfort five children if that were to happen? The big deal was that we got a new color TV to watch the launch!"

NASA children did not have normal childhoods. Speaking about it today, Kate Collins reveals a common thread in their memories: how strong their moms were. She was ten when her father, Mike, went to the moon in 1969. "I knew there were things to worry about. My best friend's dad died in the Apollo 1 fire. Airplanes crashed and missing man flyovers happened. It could happen when your dad went to work any day, not just during the spaceflights. Dad was confident and well prepared. And Mom was our firewall, so to speak. Any major anxiety was absorbed by her. We kids kept our routines as much as we could. Reporters followed us to swim class and that was weird, yet, life went on."

What added some measure of normalcy, says Kate's sister, Ann, who was seven at the time, was that all their childhood friends were also NASA kids living in the same tight-knit community of Nassau Bay, Texas, where moms ruled the roost. "I think this, as well as my young age, made things very 'normal' during those years," says Ann. She does remember her mom being nervous about her dad's flight and its effect on her and Kate. Her dad, however, with great love and patience, eased her worries by explaining to her the mechanics of the flight with models of the Saturn V, service module, lunar landing module, and the command module Mike was piloting. "He held these in his hands, arcing them through the air," Ann recalls, "so we could understand what was supposed to happen throughout the flight and landing and his orbiting alone."

Watching and listening to her dad made her feel safe. But Ann would soon be traumatized in a deceit so unforgivable that it stands out as the very worst sort of journalism practiced during the Apollo era. "The omnipresence of the press necessitated one cardinal rule that Mom strictly enforced," Ann says. "We could not open the door to anyone or go outside, even in our own backyard, without an adult. Well, one day, when the doorbell rang, I broke that rule and opened the front door by myself. Standing there were several members of the Chinese press, who very nicely handed me the most beautiful stuffed animal I had ever seen: a panda. As I stood cuddling it, they thrust a microphone and camera in my face and asked, 'Are you afraid your daddy's going to die in space?' That was the first time I was scared about the flight. I followed the rules after that."

Like Ann and Kate's mom, Dotty was vigilant about protecting her boys from the media frenzy. She knew that once Charlie retired from NASA the national spotlight would disappear, and she hoped that a more normal life might return for the family. Mostly, Dotty hoped that her Prince Charming, as she liked to call Charlie, might finally reawaken. She wanted the old Charlie back, the one who was compassionate, understanding, and romantic. They had met in Boston when Charlie was studying for his master's in aeronautics and astronautics at MIT. He was already five years out of the Naval Academy and had completed a tour in West Germany as a fighter interceptor pilot. An advanced degree from MIT was the final notch in his belt to get him into the Astronaut Corps. As a kid, growing up mostly in Lancaster, South Carolina, Charlie immersed himself in building model airplanes and watching war movies like *Flying Tigers* with John Wayne taking on the Japanese. Charlie's dad enlisted in the Navy after the Japanese attacked Pearl Harbor, and that's when Charlie felt the first stirring to serve his country. He says his dad wasn't affectionate, that he never gave him a hug or said he loved him. His aristocratic mom was the nurturer. She traced her roots back nine generations to Philemon Berry Waters, of England, whose descendants fought in the Revolutionary War. Her rule for Charlie was "never say or do anything that will cause embarrassment, and when you go out, remember who you are."

Dotty Claiborne came from an accomplished family in Atlanta. Her father was a cardiologist, and her stay-at-home mom took care of her and her sister. Dotty graduated from the University of North

Carolina at Chapel Hill with an art degree and then took a secretarial job at Harvard Business School in Boston. They met while Charlie was apartment hunting. The moment he spotted her, Charlie was instantly attracted to the blue-eyed, blond Southern belle. "A light went on in my mind and in my heart," says Charlie, who liked her sophistication and spunk. Charlie's chiseled good looks, fighter pilot bravado, and self-confidence gave him an unbeatable edge over anyone the twenty-two-year-old Dotty had ever met. In the beginning they were inseparable as Charlie studied to complete his master's. Dotty says their courtship was the most romantic time of her life. When the inevitable marriage proposal arrived, Dotty told Charlie, "'I'm going to put you first in my life. Will you put me first in your life?' His answer reassured me," says Dotty, "that I was truly the most important thing in his life."

Shortly after they got married and Charlie was accepted to test pilot school at Edwards, the needle of Charlie's attentiveness moved 180 degrees in the opposite direction. It was that fast and sudden. Like all the Eagles, his focus turned to being chosen for a moon mission, and Dotty settled into last place behind their children for his love and attention. Dotty still struggles with the whys and wherefores of his behavior but thinks some of it is traceable to the testosterone-fueled DNA of fighter pilots. Charlie came, saw, and conquered, says Dotty, and then he focused on himself and his career above everything else.

Biological anthropologist Dr. Helen Fisher, who has given numerous TED talks analyzing male-female relationships, sees common traits in the MRI scans of high-testosterone males. "They're analytical, logical, direct, decisive, calm under pressure, and able to control their emotions, which is good in an astronaut, but bad for wives looking for emotional validation and support."

Dr. Potterat says there's a similar dynamic with the SEALs, who also confront the emotional connections in their relationships. "They don't lead with emotion; they lead with intellect and sometimes within relationships there's an emotional piece that is craved on both sides that our guys sometimes struggle with."

What was completely unexpected for Dotty was that Charlie withdrew even more after his flight to the moon. Taking her cues from the women's liberation movement, Dotty thought working outside the home might bring her fulfillment. She worked at a travel agency and later helped educate low-income children at the local Head Start center. She tried throwing herself into the party circle, flirted with other men, and even tried marijuana. Nothing worked. She sank deeper into depression, and no amount of pleading for Charlie's love changed their loveless situation. She considered divorce but thought it might result in a worse fate: "The only reason I didn't divorce him is because God was speaking to me then, although I didn't recognize it was God, because the words came very clearly to my mind that there probably isn't another man out there that can love me the way I want to be loved, and so it would be like jumping out of the frying pan into the fire."

With divorce off the table, Dotty started thinking about suicide and exactly how she would do it. Her plan was to get into her car and drive headlong into the pylon of a highway overpass. She thought about it for weeks but never attempted it because members of her Episcopal church encouraged her that the "answer" lay in Jesus and giving her life to God. While Dotty had been a faithful church-goer all her life, doubt about God's existence became a dialectical battle for her. Unlike the apostle Paul, she did not have a "road to Damascus" experience where in one breakthrough moment she found Jesus. It took two months of praying (with one prayer after another being answered) before Dotty says God became real and she committed fully to the Lord. One of the first things God told her, she says, was to forgive Charlie and not put him at the center of her life but instead to put Jesus first.

Charlie says he began noticing her transformation and appreci-ated that she wasn't trying to change him anymore and became more accepting of who he was. It wasn't on the moon, like fellow moon walker Jim Irwin, where Charlie began his spiritual renewal. It was through Dotty's love and God speaking to him through a Bible study they attended. He fully acknowledged his neglect of Dotty, gave up his pursuit of money, and began devoting himself to Jesus, reading the Bible, and God's call on his life. He likes to say that his walk on the moon lasted three days but that his "walk with God is forever." Today, Dotty and Charlie are eager to tell their story of how God helped them rediscover their love for each other. The Prince Charming that Dotty found in Boston has come full lunar circle and returned to her with a committed heart.

THE LOVELLS

Jim and Marilyn at their home outside Chicago, 2015
(Courtesy of SPI-tv Media Group)

There is a scene in the movie *Apollo 13* that still makes Marilyn Lovell fidget twenty-five years after she and Jim attended the premiere. "You should have seen her when that sequence came in," Jim chuckles now, "but she sank down into her seat hoping to avoid notice." Kathleen Quinlan, the actress who plays her in the movie, is on the phone with a NASA administrator while at the same time watching real footage of ABC News science correspondent Jules Bergman giving Jim and the crew a 10 percent chance of surviving the oxygen tank explosion in the command module. Believing ABC over NASA, a highly frazzled Marilyn shouts into the phone, "Don't give me that NASA bullshit, I want to know what's happening with my husband."

When asked today if she really unloaded on NASA's official with the word *bullshit*, she gives a wry smile and says, "It just came out of me because, well, [turning to Jim] what would you call me? I'm not really a shy person. I'm a determined person. Maybe there is a little bit of the feminist in me."

At ninety-one and eighty-nine, respectively, Jim and Marilyn still sparkle with a radiant optimism that defines who they are and how they've always lived. In Marilyn's mind, if the name Rocketman could be trademarked by any one man, it would be Jim. While most kids his age played with model airplanes and electric trains, if they could afford them, Jim was making and launching miniature rockets. "I wrote to the American Rocket Society to learn about rockets," Jim says, "and I had built a little powdered rocket in high school with a friend of mine—which almost blew up."

Laughing about it now, Marilyn remembers watching the result of his efforts: "His mother and I were looking out the window watching him across the street launching this homemade rocket, and I didn't appreciate the danger as much as she did, but his mother thought he was crazy." The rocket made it off the ground, to Jim's elation, but exploded about a hundred feet up.

Jim spotted her in his junior year while working at his school cafeteria job. "I first met him in the cafeteria," remembers Marilyn, "where he was serving food, and I ended up working there myself for a while, and I was washing and drying dishes for a free meal." Drawn to her bright smile and beauty, Jim asked Marilyn, who was a freshman, to go to the prom with him. Because he was two years older, Marilyn was somewhat hesitant, and there was something else, she said. "I told him I didn't know how to dance, and he told me he didn't know how to either but that we'd learn together." Jim would bring records over to Marilyn's house so they could practice their dancing.

The prom led to a seven-year romance that endured through his studies at the Naval Academy. Like Borman, Jim was so consumed with succeeding at the academy that he wasn't able to devote as much time to Marilyn as he wanted. Marilyn had to get used to seeing Jim only on the weekends. In Annapolis she would stay in a local rooming house and would even help him type his term papers, one of which he received an A minus on: "The Development of the Liquid Fueled Rocket." Jim was way ahead of his peers and graduated at the top of his class. With graduation looming, Jim took the step he knew he was ready for. On one of Marilyn's weekend visits, he took her to the local jewelry store, got down on one knee, and asked her to marry him. Marilyn was as certain of their love as Jim was and said yes.

For the next ten years, Jim devoted his life to the Navy, where he flew missions off the Japanese coast from the USS *Shangri-La*.

His duty as a carrier pilot flying the F2H Banshee jet fighter led him to the naval air station at Patuxent River, Maryland, where he achieved top status as a test pilot. "I gave up a lot of time with him because of what he was doing, and there were eight years," Marilyn says, "where he was either on a flight or a backup. So he wasn't home very often. But I was always behind what Jim wanted to do. I didn't sit there and say, 'Well why do you want to do this,' or whatever. I was proud of what he was doing and I was right behind him."

Marilyn faced the same dispiriting crashes that Susan and Dotty did, but Jim tended to his marriage the way Bill did, always communicating and paying close attention to Marilyn's needs. He was also a master of surprise and the grand gesture. On Christmas Day 1968, as Jim was orbiting the moon on Apollo 8, he arranged for a Rolls-Royce to deliver a special package from Neiman Marcus with the following note: *To Marilyn, From the Man in the Moon.* It was attached to a mink coat.

There would be no breakdowns for Marilyn until Apollo 13: "Nothing compared to Apollo 13. I really had to be strong because I had these young children, and I put myself in a shell and I stayed in that shell, and the shell didn't break until the flight was over and he came back."

A quick death in a fighter jet, Marilyn knew, was just that: quick. You got the news within hours and you dealt with it. This was altogether different. The crew was still alive, but the explosion aboard the Odyssey put an expiration date on its oxygen and life-support systems—four days! After four days Jim, Fred Haise, and Jack Swigert would be dead. It was a mathematical certainty calculated to the last hour by NASA's flight controllers and engineers. The clock was running and Marilyn was counting the hours. For four excruciating days in April of 1970, she was trapped in her house. Apollo 13 had turned into both a global deathwatch and a prayer vigil. In St. Peter's Square, Pope Paul led fifty thousand people in prayer for the crew's safe return. Thousands more prayed at the Wailing Wall in Jerusalem. President Nixon called for a national moment of prayer. In Times Square, crowds glued themselves to the famous news ticker, which flashed continuous live updates. It was the banner headline in every day's newspapers and on nightly news programs. And then, the television crews from around the world descended on Marilyn's house. "They were like vultures," says Marilyn, who banned them from her lawn. No matter what the news people predicted, she was convinced that Jim would make it home. Jim believes they were reading each other's thoughts (almost mystically) throughout the ordeal. "She did on the ground the same thing that we did in the spacecraft," Jim said. "It's the positive energy. I mean this was a case where you had no choice but to keep going."

It's an extraordinary moment to witness between the Lovells as they exchange the look of two people who've lived in each other's hearts and minds for sixty-seven years. "I never gave up. I just had faith," Marilyn insists, "and I refused to give in to the thought that he wouldn't come back. I just wouldn't give in to that." And in the spacecraft Jim was thinking the same thing. "You had to have a positive attitude because there was no alternative. You say, 'Here's the problem, how do we solve it.' Optimism is perhaps the most basic thing a person needs to have." To which Marilyn adds the importance of prayer, something she turned to while Jim and the flight controllers worked through the engineering solutions to give them enough power to get home: "And I really have to say my faith got me through it. I mean I prayed a lot during that time, and the priest from our Episcopal church came over, and he was great moral support, and so were all the gals in the program. And then there was the news media, which had them dead the first night, and my children are listening to all these things and I couldn't think of myself because I had to be strong for them. A lot of people said I was sort of walking around in a daze at times because there was just too much going on."

After it was all over, Marilyn, understandably, unraveled. She went for therapy and bounced back quickly. Jim decided to never fly in space again. President Nixon urged him to run for the Senate as a Republican from his home state of Wisconsin, but neither he nor Marilyn found the prospect of a public life appealing. Before Jim retired from NASA in 1973, the agency sent him to Harvard Business School's Advanced Management Program, which helped in his subsequent business career. For a while he ran a tugboat company, then became president of a telecommunications company that was acquired by Centel, which kept him on as a board member and executive until 1991.

Marilyn and Jim reentered the limelight when he coauthored

his book on Apollo 13 with *Time* magazine editor Jeffrey Kluger in 1994. The success of the book and movie provided him with lifetime security and sudden demand as a public speaker. Today, Marilyn enjoys her grandchildren, Jim likes to read books on cosmology, and they walk regularly along Lake Michigan with their golden retriever, Toby. Jim still looks at the moon with longing and wishes he had walked there but says he wouldn't trade it for Apollo 13, which provided him with a fuller life and a greater appreciation of it since he almost lost his own.

Jim, Bill, and Frank have remained lifelong friends united by the magical connection of being the first three humans to have journeyed into deep space. Marilyn still attends the reunions of the astronaut wives, and it is noteworthy, although unsurprising, that after all these years the women remain closer to each other than the Eagles, who had been trained to suppress their emotions. Marilyn is happy to report that she still has, and cares for, one of her most prized possessions: the mink coat from the "Man in the Moon."

Chapter Four

LEADERSHIP LESSONS AND DOING THE IMPOSSIBLE

Wernher von Braun, center. To his right is Dr. George Mueller at the Firing Room during a test of the Saturn V rocket, 1964.

Key Lessons

➤ Find bold leaders

➤ Set a clear mission with a deadline

➤ Don't wait for perfection to make decisions

➤ Admit mistakes as they happen

Our two greatest problems are gravity and paperwork. We can lick gravity, but sometimes the paperwork is overwhelming.

Wernher von Braun

Delegate almost to the point of abdication.   *Warren Buffett*

Time, electrical power, and oxygen conspired in such a way that Apollo 13's mission clock became an hour-by-hour scorecard measuring two things that Lovell, Haise, and Swigert counted on to keep them from certain death: teamwork and leadership from NASA's engineering juggernaut on the ground. "Even though we had one crisis after another," Lovell says, "people came to the top to figure out how to get us out of those crises, like the time that we could have been poisoned by our own carbon dioxide."

The solution to the carbon dioxide problem revealed just one small piece of both the seamless teamwork and forward-thinking preparation that NASA's leaders built into the organization at every level. "I was watching TV at home when they broke into the program to say there was a problem with Apollo 13," said Ed Smylie, who was in charge of spacecraft Life Support Systems. Smylie didn't need anyone to tell him what to do next. He headed straight for the Manned Spacecraft Center (MSC), which was five minutes from his Houston home. It was his job to test and know the workings of every piece of equipment providing life support to the crew.

When he got to the MSC at 1:00 a.m., the engineering teams in Mission Control confirmed that the Odyssey was dying. The crew had moved into the lunar module (LM), Aquarius, using it as a life-boat to get home. This meant that the three astronauts had only the oxygen and power left in the lunar module to survive. Smylie, who had a reputation for being highly methodical, began calculating the consumables on the Aquarius and spotted a looming problem that could kill the crew. Some two hundred thousand miles away in the LM, Haise began the same consumables calculations as Smylie. But Haise, even though he was the LM's pilot and knew its every switch, would make one critical omission, which was understand-able given their rapidly deteriorating situation.

The Aquarius, like all the LMs, was designed to support only two astronauts—the two descending to the lunar surface, while the pilot in the command module stayed behind orbiting above. Now it had to support three, and Haise forgot to factor command module pilot Swigert into his calculations. Not Smylie. The unas-suming Mississippi farm boy, with an engineering degree from Mississippi State University and a master's from MIT, knew that the added carbon dioxide emissions from Swigert would soon over-whelm the two lithium hydroxide canisters recycling the CO_2 in the Aquarius. A solution began forming in Smylie's head. He would use the two recycling canisters from the command module to soak up the extra CO_2 load from Swigert. Only one problem had to be solved—the male-female connectors on the Odyssey's canisters were incompatible with those on the Aquarius. To connect the two, the engineers had to figure out a way to fit a square peg in a round hole. Smylie says the movie made it appear that he alone figured out the solution when, in fact, his notes show that sixty people and forty different contractors worked together nonstop for twenty-four hours to jury-rig and test a contraption that consisted of duct tape, plastic bags, tubing, socks, and cardboard. The crew successfully stitched it all together just as the CO_2 levels were reaching the danger zone.

When the crisis was over, President Nixon presented Smylie and his team with the Presidential Medal of Freedom for solving the carbon dioxide crisis: "Had that not occurred these men would not have gotten back," Nixon said. "That is only one example to prove the magnificent teamwork of the whole group."

NASA itself has described Apollo 13 as its finest hour. Among other things, it produced a management axiom that is still a favorite among business leaders and is a go-to motivational phrase in crisis situations: "Failure is not an option," which originated with flight director Gene Kranz, who told his assembled team that the Apollo 13 crew would not die on their watch.

If there was one statistic that captured the scale and complexity of the moon shot, and produced night terrors for NASA's management, it was this: the Saturn V moon rocket, and the command and lunar modules it carried, contained 5.6 million parts. Consider, also, that some four hundred thousand engineers, scientists, and technicians, and twenty thousand contractors had a hand in it. Those daunting organizational dynamics produced an unnerving mathematical calculation for management and the Eagles, which would come to inform all their management decisions. It came down to this: assuming a 99.9 percent reliability rate in all phases of assembly, that meant 5,600 parts could potentially fail—any one of which could result in a tragic outcome. It was not a theoretical exercise. The dreaded 0.1 percent failure risk, as we've seen, emerged in the fatal Apollo 1 fire and in the faulty valve on Apollo 13 that caused oxygen tank number two to explode.

Looming over NASA's sprawling national effort was the pressure of a sacrosanct eight-year deadline, mandated by President Kennedy, to land on the moon before 1970. Jeff Bezos, who scaled Amazon into the giant it is, believes NASA's feat was one of the greatest *organizational* achievements in human history. "We as a civilization, we as humanity, pulled that Moon landing way forward—out of sequence—from where it actually should have been. It was a gigantic effort which in many ways should have been impossible," says Bezos, who points out that today's smartphones have more computational power than the Apollo spacecraft. "They had to invent a lot of what today would be foundational operations research kinds of ideas…it's amazing that we did it in 1969."

For those who were paying attention, *Science* magazine, as early as 1968, recognized in a lengthy article that NASA's leadership methods during Apollo would prove to be as important in the long run as its technological achievements:

It may turn out that the space program's most valuable spin-off of all will be human rather than technological: better knowledge of how to plan, coordinate, and monitor the multitudinous and varied activities of the organizations required to accomplish great social undertakings.

John Aaron, who would continue at NASA through the space shuttle program, says he echoed that point years later at a technology conference examining the main scientific components that got us to the moon: "Of course the propulsion guys would talk about engines, computer guys would talk about computers and so forth. And so at the end of that little discussion I said, 'Well excuse me, but I think we need to recognize that there's one other technology that was very critical to our success and that was the management technology,'" said Aaron. "Where else could you put a management team together overseeing four hundred thousand people that could build all these thousands and millions of parts across the country, and do it on schedule, and deliver it all to the launch site and make it work?"

Neil Armstrong would pay tribute to that human effort as he flew home from his historic walk on the moon: "We would like to give a special thanks to all those Americans who built the spacecraft, who did the construction, design, the tests, and put their hearts and all their abilities into those crafts. To those people tonight, we give a special thank-you."

To analyze the management methods used to achieve the moon

shot, NASA's History Division gathered the six key executives over-seeing its success in a daylong symposium held in 1989. Little has been written about it, and only Dr. Chris Kraft (age ninety-five), who was NASA's first flight director, is still living. Here are the ten management principles the conferees identified that would be recognizable to any CEO today running either a large or small organization:

> Have a clear mission
> Find leaders with vision
> Hire the best people
> Have sufficient funding
> Delegate authority
> Foster teamwork and trust
> Own mistakes (and admit to them in real-time)
> Engage in open communication and coordination
> Ensure quality control
> Take calculated risks and have moral courage

Charlie Duke, who like many of the Eagles gives speeches on leadership, describes how effectively NASA's management practices and principles were put into action: "Well, first, NASA picked the right people. Then the right people that I interfaced with delegated authority, and they stayed focused on their mission, and they listened to their subordinates and the engineers in their area of responsibility. And they had an open door, and they were knowledgeable enough in their area of expertise to figure out, 'Yeah, we can do that, and no, we can't do that.' So it was knowledge on their part, it was the ability to delegate authority and keep the whole team focused on their goal, and that was to make sure our part, or their part, was delivered on schedule—and it worked."

A CLEAR MISSION & VISIONARY LEADERSHIP

Today, we owe the term *moon shot* to President Kennedy's decision to put a man on the moon. It is defined by WhatIs.com, one of the glossary websites for Silicon Valley, as "an ambitious, exploratory and ground-breaking project undertaken without any expectation of near-term profitability or benefit and also, perhaps, without a full investigation of potential risks and benefits."

"Was Kennedy a visionary, was he a dreamer, was he politically astute? The chances are yes, he was probably all three," said Gene Cernan.

When President Kennedy stepped in front of Congress on May 25, 1961 (and later at Rice University), he unveiled a mission that was breathtaking for its magnitude:

We choose to go to the Moon. We choose to go to the Moon in this decade and do the other things, not because they are easy, but because they are hard, because that goal will serve

to organize and measure the best of our energies and skills,
because that challenge is one that we are willing to accept,
one we are unwilling to postpone, and one which we intend
to win, and the others, too.

It was masterful in its articulation of a clear vision, with a clear
goal, and a clear deadline. No president had ever challenged the
nation with such a wildly improbable undertaking. "It was beautiful in its simplicity," says Mike Collins. "Do what? Moon! When?
End of the decade."

I believe we possess all the resources and talents necessary.
But the facts of the matter are that we have never made the
national decisions or marshalled the national resources
required for such leadership. We have never specified long-
range goals on an urgent time schedule or managed our
resources and our time so as to insure their fulfillment.
But in a very real sense it will not be one man going to the
Moon—if we make this judgment affirmatively, it will be an
entire nation. For all of us must work to put him there.

By enlisting all Americans, Kennedy signaled the immense
scope and scale of the effort. He then put his political capital on
the line by attaching real numbers to a project that was far from a
sure bet:

Let it be clear that I am asking the Congress and the country to accept a firm commitment to a new course of action, a course which will last for many years and carry very heavy costs: 531 million dollars in fiscal '62—an estimated seven to nine billion dollars additional over the next five years. If we are to go only half way, or reduce our sights in the face of difficulty, in my judgment it would be better not to go at all.

Until Kennedy's speech, America's total flight time in space was the fifteen minutes Al Shepard spent in a suborbital hop on Freedom 7. The Soviets, by comparison, were beating America every step of the way. It started with the launch of the world's first satellite, Sputnik, in 1957, followed by Yuri Gagarin grabbing the historic headline of being the first human to orbit the Earth in 1961. America's political and business leaders were legitimately starting to panic that the Soviets were surpassing the United States in the sciences and the new frontier of space. It was the height of the Cold War, and the Russians propagandized their success as proof that Communism was a superior political and economic system to America's capitalist democracy. On top of that, Kennedy had just suffered his first humiliating foreign policy disaster at the Bay of Pigs, in Cuba, when 1,400 American-trained Cubans were defeated by troops of the newly installed Communist leader, Fidel Castro. It was a pivotal moment in world affairs, and Congress was sufficiently scared and motivated to fund Kennedy's request. "Politically it was about beating the Russians," said Ed Mitchell, "but those of us with a science bent, a curious bent, knew it was much more than that."

In 1958, NASA scientists were already quietly studying the mechanics of getting to the moon. "We knew what had to be done," said Dr. Max Faget, who was the head of engineering. "How to do it in 10 years had never been addressed before the announcement was made." Kennedy had not just wildly issued the challenge. Before making his pitch to the nation, he consulted with Dr. Robert Gilruth, NASA's first director of the Manned Spacecraft Center. "Can it be done?" Kennedy asked Gilruth. "I told him that I thought that maybe we could go to the Moon, but I wasn't sure that we could. And there were a lot of unknowns that we would have to uncover before we were sure. And he said well, let's go ahead and say we can do it in a decade. And we will do the best we can."

Kennedy had just laid the predicate for a teachable moment in leadership and moral courage. He was a realistic idealist who appreciated that calculated risks could lead to greatness. Kennedy had explored risk and moral courage in his 1957 Pulitzer Prize–winning book, *Profiles in Courage*, in which he honored public servants willing to risk their careers for worthy causes with uncertain outcomes. Kennedy was prepared to make the bet that NASA and America's industrial establishment could work together in a way that had never been attempted before.

"Kennedy gave us the key imperatives to meet his goal," says Borman. "Number one we had a clear mission to get to the Moon and back in the decade of the sixties. And number two, we had the support of the Congress; we had the support of the president; and we had the support of the people—there was almost unanimous support, particularly in the beginning—for the space program; and then we had people that were willing to dedicate everything to make it go. It was in my estimation almost like a maximum war effort. And it made NASA, quite frankly, a wonderful place to work."

HIRING & FUNDING

With the goal set, now it was time to hire the right people. Given the extended time horizon of the moon shot—eight-plus years—Kennedy's political radar told him that NASA would face constant funding battles and potentially fickle public support. To fight against both, Kennedy hired an old Washington insider, James Webb, to lead NASA's efforts. Webb was an ex-Marine-pilot with a big, firm handshake attached to a talent for getting what he wanted. His résumé included a law degree from George Washington University and an eight-year run as the vice president of Sperry Gyroscope Company, which provided the Department of Defense with navigation equipment and airborne radar systems during World War II. He had business experience and connections to many of the key contractors who would later build the spacecraft for the moon landing. But what really qualified Webb to run NASA, in Kennedy's estimation, was his subsequent job in the Truman administration as director of the Bureau of the Budget. Webb understood budget battles. He understood the politics of how Congress allocated money. He knew which strings to pull and whose districts stood to gain the most from NASA contracts. The fifty-five-year-old Webb proved to be an inspired choice.

Webb made two pivotal decisions right out of the gate that would ensure NASA's success—each having to do with money and politics. First, he told Kennedy something that any venture capitalist who has funded a high-tech startup would understand. In his speech to Congress, Kennedy estimated the cost of the moon shot at between $7 billion to $9 billion. Webb told Kennedy to double it to $20 billion. He had seen too many government programs (and private businesses) suffer slow deaths from inadequate funding estimates, and he wanted NASA to have the largest cushion possible to succeed.

"Webb's decision to double the budget estimates guaranteed NASA's survival," says Schmitt, who later served as a US senator from New Mexico. "As a result, NASA started out the Apollo program with 100 percent management reserve. And that meant that any time that they met a problem, or an unexpected problem, they had the resources to press on and maintain schedule."

Second, Webb wanted a long-term insurance policy that vested the politicians controlling the money. He knew just where to find it: the House Appropriations Subcommittee, which oversaw NASA's budget allocations. Its chair was Congressman Albert Thomas from the increasingly powerful state of Texas, which helped Kennedy win the 1960 election. It also happened to be the home state of Vice President Lyndon B. Johnson, whom Kennedy had picked to oversee the space program. Webb was about to hand Johnson's and Thomas's constituents a Texas-size gusher, which would keep them motivated for the long run. By 1962, NASA had outgrown its research center in Langley, Virginia, and what's more, the state's senator, Harry Byrd, had done little financially for the space program. Webb needed a new center to house the thousands of new engineers and technicians it would take for the moon landing (thirty-five thousand direct NASA hires). It would mean new jobs, new construction, and innumerable spin-off benefits, so Webb picked a city for the new center that just happened to be in Thomas's voting district: Houston. Thomas happily swallowed the bait. He and Johnson made sure that NASA got the funding it needed, and it is no coincidence that today its headquarters is called the Johnson Space Center.

"Webb knew government and how to get funding," says Borman. "Technically he was not that sharp because basically he was a lawyer and a politician, but he appointed underneath him really brilliant people, and he let them do their jobs and didn't get in their way. It wasn't a top-down management structure, a valuable lesson I learned, which I used when I became president of Eastern Airlines."

With funding now assured, Webb needed the best scientific brains in the country, which meant offering salaries competitive with private industry. To make it happen, Kennedy and Webb saw to it that Section 203(b)(2) of the space act provided NASA's key officers and scientific talent with a new category of "supergrade" compensation as part of the larger 1962 Salary Reform Act. It made a difference. From TRW, which built Pioneer 1, NASA's first unmanned spacecraft, in 1958, Webb tapped one of its top directors, George Mueller. Mueller's job was to oversee NASA's new Houston center and its other key centers located in Huntsville, Alabama, where the Saturn V rocket was built, and Cape Canaveral, Florida, where the launches took place.

Mueller was a brilliant taskmaster born in St. Louis in 1918 to a father who was an electrician and mother who had been a secretary. As early as age twelve, he was building his own radios. He wound up getting a television fellowship from RCA to study at Purdue University, where he earned his MS in electrical engineering, followed by a PhD from the Ohio State University. At Bell Labs, the nation's foremost research laboratory at the time, he helped design airborne radar systems. Mueller was an impresario at managing large research organizations doing big technology projects.

When he arrived at NASA, he sought engineers who had at least five to ten years of experience in manufacturing, design, and development. They came primarily from the United States, Canada, and the United Kingdom, and most were in their mid- to late thirties. But in Houston's Mission Control (MC) it was a different story. MC was the public-facing nerve center for all spaceflight operations. The memorable televised launch countdowns emanated from the flight controllers in MC. They talked to the astronauts in space, ran the computer programs, telemetry, coordinated reentry, splash-downs, and capsule recovery by the Navy in the Pacific Ocean, and much more.

Flying in space was new. "Nobody really knew how to go to the Moon," said David Scott, the seventh moon walker. "It was a lot on paper, and we didn't know how to do things, and we didn't know how things would work; it was just a matter of putting them together, making it work, and then correcting deficiencies, and as pilots-astronauts we participated in all of these things along with management and the engineers."

At Mission Control there were no organizational or operational precedents for any of it and no pool of experienced talent to draw from. Dr. Kraft created it all from scratch, and because it was new, Kraft wanted engineers with immense energy who were not set in their ways. "I looked for people right out of college. That is where I wanted them from....And I want you to know that the average age of my organization in 1969 was 26," Kraft told his fellow conferees in 1989.

"That was a really smart move by Kraft," says Schmitt, "hiring fresh college graduates. They were young, all in their twenties, had stamina, courage, as well as the motivation and the patriotism to continue to work sixteen-hour days, eight-day weeks. And that was a very, very important part of meeting Kennedy's deadline."

"When I went to NASA I didn't know anything," says Aaron. "I learned everything at NASA, just by probing my curiosity, and they made sure you performed, and it was very clear what the expectations were. And I think from that process, a very important principle that I learned as a manager later, is never underestimate what you can get people to do. Always put the expectation out there of doing a little better than they're performing. And you'd get surprised, they'll do it."

DELEGATE & COMMUNICATE

"Another thing that I think was extraordinary, and this was throughout the whole manned space flight program, was how things were delegated down," said Bill Tindall, who helped perfect the docking method (orbital rendezvous) of the command and lunar modules. "I mean, NASA responsibilities were delegated to the people and they, who didn't know how to do these things, were expected to go find out how to do it and do it. And that is what they did. It was just so much fun to watch these young people take on these terribly challenging jobs and do them."

"I think we had a group that was not only learning, but was very, very capable," said Mueller. "And truthfully, that was the secret of success in NASA, the capability of the individuals involved in all of the centers because we had some tremendous people down at the Cape, as well as in Huntsville. I think that one other thing that was instrumental was the fact that we were able to work quite openly with our contractors."

"And the other thing you have to realize," said Dr. Kraft, "is there was a tremendous feeling of openness among our organizations." Kraft himself had the deepest roots at NASA, having started at its predecessor agency, the National Advisory Committee for Aeronautics (NACA), straight from Virginia Tech. He spent ten years in aeronautical research before joining the Space Task Group, which was tasked with laying the groundwork for putting a man into space. "We grew up telling each other we were making mistakes when we made them. And that is how we learned. It was extremely important for us to say the mistakes we made as we made them because that helped us to grow. And that feeling was very much a part of our organization."

To help eliminate and catch mistakes early, Mueller created what modern management experts call a *flat organization*, in which there are the fewest layers of management possible separating staff from executives. This was especially important for Wernher von Braun, who was building the Saturn V rocket. Ferreting out design errors was his top priority. To help prevent them, von Braun created a management tool for his team called the Monday Notes, encouraging engineers at the ground level to report potential problems early in the design process.

He limited the notes to one page, and managers at all levels were required to submit them each Monday morning for extensive review.

"There weren't many layers of bureaucracy between the engineer and von Braun, and those layers could be penetrated if someone at the bottom felt strongly enough about a developing problem," says Borman, who worked closely with von Braun on an "emergency detection system" in the event of a launch abort.

"Apollo had a very minimally layered decision-making system, and a good idea at any level, if it was a good idea, could work very rapidly up the chain," says Harrison Schmitt, who met with President Trump in 2017 about returning to the moon. "In fact, the chain probably only had one link in it before it got to a decision maker."

To ensure proper communication between the centers, Mueller further created something called the five-box management structure, which clearly defined everyone's responsibilities:

1. Headquarters
2. Program control
3. Systems engineering
4. Flight operations
5. Reliability and quality

Mueller said that when he created the boxes, he inserted a program management structure in parallel with the functional structure of the centers:

You had to set up the interfaces within that system clearly and fix them so that everyone understood what those interfaces were. And you had to have program control, so somebody was keeping track of scheduling dollars and what the implications were. And finally, you had to have someone who was worried about when you had all of this put together, if it will fly and how you will fly it, so we had to have an operations box. And we duplicated this down through the structure in such a way that there were communications between like disciplines so that you could be sure that there was the right set of information flowing up and down the chain in order to be able to make decisions and to follow the program and be sure that everybody was in sync.

OWNING MISTAKES & ENSURING QUALITY CONTROL

As is inevitably the case in any organization, not everyone was as "in sync" as Mueller had hoped. While Mueller and his engineering teams strove for zero defects, the information flow between managers (and managers and contractors) started to break down in the first stages of developing the Apollo spacecraft, which began in 1963. The five-box management structure worked well for the six Mercury and ten Gemini flights, which flew between 1961 and 1966. But the pressure of meeting the 1969 deadline to land on the moon short-circuited NASA's normal "reliability and "quality" controls that had served it so well.

"What we did in the early days," says Mike Collins, "was take the overall spacecraft and divide it up like a pie. We sliced that pie up into ten or fifteen different pieces, and we handed each slice to one of the astronauts and said, 'This is yours, we want you to learn that slice.'" Recruiting the astronauts in the design process was astute since they might spot flaws that weren't always obvious to engineers who'd never flown in space, and management would listen to their suggestions. The tragic exception to this was the case of Apollo 1. In what may be the most hauntingly prescient photo of the Apollo era, the three astronauts assigned to Apollo 1 took a grainy photo of themselves with heads bent in prayer over a model of the command module.

From left to right: Ed White, Gus Grissom, and Roger Chaffee, 1967

Hoping for a higher power to intervene, Grissom, Chaffee, and White put an inscription on the photo, in semi-jest, that read: "It isn't that we don't trust you, Joe, but this time we've decided to go over your head."

The Joe referred to was Joe Shea, who had been hired by Mueller from his old firm, TRW, where Shea also worked on ballistic missile systems. Shea received his PhD in engineering mechanics from the University of Michigan and was considered a superb engineer, but Kraft complained that he was a lousy manager. Shea's organizational challenge was to ensure that everything coming off the production line from hundreds of contractors and subcontractors fit together and worked according to the specifications for the three pieces of the Apollo spacecraft: the command module (CM), service module (SM), and lunar module (LM). The sprawling complexity of it, multiplied by the urgent timeline, eventually led to Shea's nervous collapse, particularly after the fingers started pointing to him for the Apollo 1 fire.

"The animosity between my people and Shea's was intense," Kraft reported. Kraft was furious because neither Shea nor Apollo's primary contractor, North American Aviation, were sharing systems information and schematic drawings with his flight controllers... unlike McDonnell Douglas, which had built the Mercury and Gemini spacecraft with few glitches. By the time Kraft's team got access to the schematics and began interfacing with Shea's group, the requested design changes were numbering in the hundreds. North American blamed the changes for production delays and cost overruns. And worse, they said, NASA failed to heed their advice in two critical areas that could lead to a catastrophic event. The first was running a ground test using 100 percent pure oxygen in the space capsule, which could ignite a fire should a spark occur. The second admonition was that NASA *not* change the explosive-bolt hatch design of the Mercury and Gemini capsules, which allowed the astronauts to get out quickly in case of an emergency. NASA ignored both warnings, reasoning that the risk of a ground-test fire was infinitesimal. However, its engineers thought that the risk was high (on a long flight to the moon) of the hatch blowing accidentally because of the explosive bolts.

North American was correct about the oxygen and the hatch design, but it was their own shoddy workmanship that triggered the fire catastrophe they had warned against. Grissom, Chaffee, and White had complained repeatedly to Shea about exposed and frayed wiring in the capsule that they were testing (which was later found to be the likely cause of the lethal spark).

"He was unquestionably a brilliant scientist," Frank Borman would later say of Shea, "but also a poor administrator who simply let North American's design mistakes pile up like unnoticed garbage."

To make matters worse, Shea made the fateful error of reporting the problems with North American only to Mueller, who was unaware that Dr. Gilruth, the man in the first position to resolve them, was left in the dark. It was the worst kind of communication failure possible. The final management sin was Shea's failure to act on the questionable workmanship before it was too late for the Apollo 1 crew. Frayed wires, 100 percent oxygen, and an inward-opening hatch that was a struggle to open doomed the three astronauts to an inescapable death.

Shaken to the core by the crew's loss, Kranz called a meeting of his flight controllers in the auditorium, which expanded to include spacecraft contractors and civil servants. His speech became another of those teachable moments when a leader finds the right words to reassure a team facing both grief and guilt over the death of men who entrusted them with their lives:

> From this day forward, Flight Control will be known by two words: Tough and Competent. *Tough* means we are forever accountable for what we do or what we fail to do. We will never again compromise our responsibilities. Every time we walk into Mission Control we will know what we stand for.
>
> *Competent* means we will never take anything for granted. We will never be found short in our knowledge and in our skills. Mission Control will be perfect.
>
> When you leave this meeting today you will go to your office and the first thing you will do there is to write "Tough and Competent" on your blackboards. It will *never* be erased. Each day when you enter the room, these words will remind you of the price paid by Grissom, White, and Chaffee. These words are the price of admission to the ranks of Mission Control.

The Apollo 1 fire led to a four-month internal NASA investigation, followed by congressional hearings in which Webb, Mueller, and other top executives had to confront their oversight failures leading to the astronauts' deaths. Joe Shea became so unglued that, at a presentation he was giving investigating the fire, he started babbling incoherently while his colleagues looked on in discomfiting horror. He would later be reassigned to a desk job in Washington. North American fired Harrison Storms, who was in charge of managing the design and production of the command and service modules. James Webb would retire one year later, honorably, when the Nixon administration came into office, but his legacy had taken a slight hit.

"When we had the fire," Kraft said, "I think we took a step back and said okay, what are these lessons that we have learned from Mercury and Gemini. What lessons have we learned from this horrible tragedy. And now let's pump that back and be doubly sure that we are going to do it right the next time. And I think that that fact right there is what allowed us to get Apollo done in the '60s." Owen Morris, who was chief engineer of the lunar module, added to Kraft's assessment, saying, "As a result of the accident, the fire, the procedures were really tightened up."

"It wasn't just that we fixed the fire, we fixed everything else we could find that had any possibility of being fixed," concluded Mueller.

In total, 125 things needed fixing, and NASA would spend

the next twenty-two months with its contractors redesigning the Apollo spacecraft. To make sure the job was done right this time, Webb replaced Shea with George Low, who was a close confidant of Gilruth's and was highly respected for his administrative skills. Low had joined NASA in 1958, so his agency experience was long and deep. One of Low's first moves was to set up regular visits to NASA's contractors: "He had a Configuration Control Board meeting," said Kraft, "and once a month it was held at Grumman and at Rockwell. We would fly from here [Houston] to Grumman, have an all-day meeting, get on an airplane, fly all night, have the same meeting in California with Rockwell and fly home.... So everybody was familiar with what was going on in all of those contracts as far as we were concerned."

With the Eagles, Low decided it was important to hold weekly meetings giving them a full hearing for any safety or other issues they identified. He wanted no repeats of Grissom's ignored warnings. "If a problem surfaced," says Duke, "and we spotted a safety or flight problem that we wanted to fix quickly, or we had an idea about how to simplify some part of the operation, and you could convince Low that you were right, he'd make a decision and say okay, do it."

Borman, for example, found a big problem with the flight controls of the redesigned Apollo command module, which he was testing at North American's factory: "I got into the Apollo simulator and tried out the controls. I pulled back on the stick and the nose went down. I pushed forward and the nose went up. Exactly the opposite of flying an airplane." Borman called over the engineer in charge and told him he had the polarity reversed, and that neither he nor any other astronaut would fly it unless he fixed it. When the North American engineer challenged him, Borman says, "I called the people at Houston, told them what a dumb thing North American had done and got an official order to keep the stick oriented properly." NASA listened. The problem was fixed.

By late 1968, Webb, Low, Mueller, and Kraft had everyone back on track, and the Apollo command module was ready to fly. Their next move would test the nerves of the entire organization.

CALCULATED RISK

"There is a time to be conservative and a time to be bold. And judgment, good judgment, tells you when to do it," said Dr. Faget, referring to the decision to send Apollo 8 to the moon. It was "the single greatest gamble in space flight since and then," wrote Al Shepard and Deke Slayton in a book they coauthored as heads of the Astronaut Office.

"We hear from the CIA that the Russians are going to send a spacecraft around the moon, or with a person in it and upstage us," said Alan Bean.

"If they orbit the moon before we land on the moon, then

they've gotten there first," said Gene Cernan, the last human being to walk on the lunar surface.

And NASA was not going to let that happen. "We changed our plans on Apollo 8. We changed the mission from an earth orbital type to a flight to the Moon," says Lovell. "And it was a bold move. It had some risky aspects to it, but it was a time when we made bold moves."

"We spent a great deal of time, energy, and effort being sure that we understood the risks," said Mueller. "So it wasn't that we were just boldly marching out where angels feared to tread. We really understood what our system was."

"We knew what we were doing," reiterated Tindall. "We had procedures laid out, we tried to imagine every single conceivable failure that could occur and what we were going to do about them." The three Eagles that NASA tapped for the greatest voyage in human history, Borman, Lovell, and Anders, appreciated the risks involved, but they had come to trust the entire NASA team and the work it had done. Six years had passed since President Kennedy inspired the nation with his vision to be first to the moon . . . "and do it right," he said, "and do it first before this decade is out—" What frequently gets dropped from Kennedy's oft-quoted sentence above are the five words following the dash—*then we must be bold*. The Eagles accepted boldness and calculated risk as two of the funda- mental ingredients required of great human endeavors.

Thanks to newly invented technologies, and driven by the

collective management genius of a select group, Borman, Lovell, and Anders began orbiting the moon on Christmas Eve 1968. They were changed by a view no humans had seen before. Their interplanetary observations, broadcast on live television, would turn them into instant messengers for a new enlightenment whose impact is still being felt today.

Chapter Five

CHANGED BY A VIEW:
THE COSMIC LIGHTHOUSE

Alan Bean in his art studio, 2017 *(Courtesy of Basil Hero)*

KEY LESSONS

- ➤ Reframe your perception of Earth
- ➤ Remember we are one race, the human race, sharing the same planet
- ➤ Have greater empathy for our neighbors' plight
- ➤ Rethink your approach to conflict from a planetary perspective

We were looking at things that human beings had never seen
before.   *Harrison Schmitt, Apollo 17*

We'd trained for everything except for *Earthrise,* which really
caught my attention and basically changed my life.

Bill Anders, Apollo 8

The most awesome thing, and the thing that really got my
imagination, was looking at the universe out there.

Al Worden, Apollo 15

To understand the Eagles, one has to start with the voyage of
Apollo 8, which the *New York Times* reported as "the greatest adventure in human history." Its magnitude and impact on the Eagles
cannot be overstated. On the morning of the flight, Anders, a practicing Catholic, took communion from his childhood priest. Five
days later he would return an agnostic, questioning everything the
church had taught him.

On Christmas Eve 1968, one billion people joined in the
largest television audience (at that time) for a single event: Apollo
8's live television broadcast of the first close-up pictures of the
moon.

The anticipation had been building for years, particularly in
schools around the country where they would roll TV sets into the
hallways for each launch of the Mercury and Gemini spacecraft.
Like a sacramental ritual, students would find themselves joining

in the T-Minus countdowns shouting out "Ten, nine, eight, seven," right up until "zero," and "liftoff" when the erupting fireball would fan out like some monstrous peacock lifting the rocket in slow motion, teasing it to punch a hole in the sky. For school kids of a certain age, there was nothing cooler!

Seven years after Shepard's short flight, the first manned moon shot had finally arrived. His friends Borman, Lovell, and Anders were circling the moon 230,000 miles from Earth. Of the billions of eyes that have stared at it since humans evolved, theirs were the first to see behind the far side of the moon, the first to see the details of its ancient craters, and the first to hover over its surface like Olympian gods sizing up its uninviting domain.

NASA gave the world a front-row seat for the lunar voyage thanks to the revolution in global communications it helped pioneer through its first-generation computers and satellites, thereby laying the foundation for today's interconnected smartphone world in which life and major events unfold in real time.

Until Apollo 8, historians relied on the diaries of the great explorers to memorialize their new discoveries. In 1492 there were no live cameras on the *Santa Maria* when Christopher Columbus stepped foot on the New World, none when Magellan circumnavigated the globe, and none broadcasting from Charles Lindbergh's single-engine plane during his thirty-three-hour solo crossing of the Atlantic Ocean in 1927.

But now, Earth's rainbow of humanity, peering through television screens on every continent, was about to see exactly what the

Apollo 8 crew were seeing…as they were seeing it…for the first time. Even if the images were fuzzy, the storied journey was unfolding live—suddenly, Frank Borman's voice began crackling through the vacuum of space as he, Anders, and Lovell took turns describing the moon's vistas pulsing across people's TV sets.

FB: This is Apollo 8, coming to you live from the moon…. the moon is a different thing to each one of us…. Each one carries his own impression of what he's seen today. I know my impression is that it's a vast, lonely, forbidding-type existence, or expanse of nothing, that looks rather like clouds and clouds of pumice stone, and it certainly would not appear to be a very inviting place to live or work. Jim, what have you thought most about?

JL: Well, Frank, my thoughts are very similar. The vast loneliness up here of the moon is awe inspiring, and it makes you realize just what you have back there on earth. The earth from here is a grand oasis in the big vastness of space.

FB: Bill, what do you think?

BA: I think the thing that impressed me the most was the lunar sunrises and sunsets. These in particular bring out the stark nature of the terrain, and the long shadows really bring out the relief that is here and hard to see at this very bright surface that we're going over right now.

Photographing the moon to find suitable locations for Armstrong's landing was essential to their mission, but on their fourth orbit, the three laconic astronauts (trained with just-the-facts emotions) saw something rising above the lunar horizon that, like a cosmic thunderclap, shattered their worldview and altered their perception of the universe and humanity's place in it.

All three men, uttering *my Gods* and *wows*, forgot momentarily about the moonscape below and scrambled for their cameras and color film to capture it. It was not some mystical panorama out of *Star Trek*, which first aired on television in 1966, nor a black, Monolith-like enigma from *2001: A Space Odyssey*, whose movie premiere they attended before their flight.

It was something more divinely counterintuitive.

It was *Earth*. Our mother ship.

Until Apollo 8's moon voyage, civilization's observational vantage point had always been earthbound looking up, never the other way around. Suddenly the view was reversed. Lovell had already registered his wonderment during the Christmas Eve broadcast, revealing a preternatural stirring (present to this day) when he marveled at Earth's galactic exceptionalism. Reflecting today, Anders captures the grand paradox of the mission, saying, "We went to explore the Moon, but discovered the Earth instead."

Circling the moon generated a Galilean moment in reverse, a revelation as momentous as Copernicus's as they focused their Hasselblad cameras homeward. As we all learned from Copernicus, it is the sun, not the Earth, at the center of our solar system, but there was something else the astronauts educed—something disorienting, something totally unexpected. And all three men drew the same conclusion about it as they focused on the blue water planet encased in the black shroud of space: small and fragile.

The notion of the Earth as *small* and *fragile* had never figured into man's thinking or appeared before in letters of philosophy and science. While the turbulent 1960s reignited the nineteenth century's concerns over industrial pollution and overpopulation (and the newer threat of nuclear annihilation), the *whole* Earth itself appeared vast and indestructible. Whether it was topping the highest mountain or crossing an untraversed ocean, Earth still produced awe and a healthy trepidation for centuries of explorers who risked their lives to probe its uncharted reaches.

This reverential perspective remained true even for the first men in space—the low-earth-orbiting Mercury and Gemini astronauts, such as John Glenn (and even the space shuttle crews), who typically orbited 330 miles above the planet—too low to see it as a complete sphere. As with today's International Space Station (ISS) astronauts, what they saw was the entire curvature of the Earth and its wholly observable continents and oceans. To them, the home planet looked majestic, and by circling it every ninety minutes, it became the focal point of all their senses.

For Borman, Lovell, and Anders, precisely the opposite perception of Earth occurred because only the Eagles have seen Earth reduced to the size of a marble. In order for that to happen, they needed to escape the Earth's gravitational pull, which meant their spacecraft had to achieve what's known as escape velocity—a speed about ten times that of a bullet from a high-performance rifle— 23,226 miles per hour in order to break free from the planet.

Only the Eagles have experienced escape velocity.

"At that velocity, I could see the Earth start to shrink," Lovell explains. "I put my thumb up to the window and thought behind my thumb were 3.5 billion people on that small planet that I can hide completely. And I thought to myself how insignificant we really all are."

Earthlings saw Lovell's Earth-behind-the-thumb perspective three days after Apollo 8 returned home, when NASA published the color photo taken by Anders during their fourth pass around the moon. Its forerunner was a crude black-and-white photo of Earth taken in 1966 by the unmanned Lunar Orbiter 1, whose faint image barely registered outside the scientific community.

By comparison, *Earthrise*, as it eventually came to be known, was a pristine photographic jewel, which all at once widened the eyes of humanity when it saw precisely what the planet looked like from another world. Humans had never seen their home from deep space, in color…let alone with such clarity in a photo taken by human hands.

"I think that seeing the Earth rise from the moon that Christmas Eve was a remarkable experience, said Borman. "All three of us were awed by it immediately…it was the only thing in the universe that had any color, and I was totally taken by it."

August 23, 1966, by NASA's Lunar Orbiter 1

Earth's satellite, however, was starting to sound a lot less romantic. "The moon looks like a battlefield. It's ugly!" Anders said as he pointed to pictures of it in his study. "It was nothing but holes and holes upon holes, and after three orbits I thought there's nothing new to see, it's boring! And when we saw the Earth on our fourth orbit, I thought, *Wow!* That's more important in a sense than the moon."

Earthrise, taken by Bill Anders, Christmas Eve 1968

Mike Collins would have the same reaction one year later during his lunar orbit: "When the sun is shining on the surface at a very shallow angle, the craters cast long shadows, and the moon's surface seems very inhospitable, forbidding almost. I did not sense any great invitation on the part of the moon for us to come into its domain. I sensed almost a hostile place, a scary place."

Even though they were the first Apollo crew to leave Earth's orbit, this inversion of expectations was identical for each of the eight crews who followed. No matter how many times Borman, Lovell, Anders, and those after them described the experience of seeing Earth from the moon, each new crew came to its lunar nest as newborns.

And it was easy to see why. When President Kennedy issued his lunar challenge to the nation in 1961, the moon, that mythical object of magical thinking and infinite enchantment, was all the astronauts could think about. The luminous moon was the bull's-eye focus of their many years of training—of surveying it, photographing it, determining how to *get* to it, *land* on it, and *walk* on it, and which areas to explore for geological significance. When they finally did touch its surface, the moon walkers (as opposed to those who merely orbited it) discovered a haunting beauty and a spooky sequestration:

"When you land on the moon, and you stop, and you get out, nobody's out there—this little lunar module—and then the two of you," said Alan Bean. "You're it, on this whole big place, and that's a weird feeling, it's a weird feeling to be two people and that's it."

"I think the feeling I had the whole time was the feeling of awe," said Duke. "The moon was the most spectacularly beautiful desert you can ever imagine, unspoiled, untouched. It had a vibrancy about it, and the contrast between the moon and the black sky was so vivid, and it just made this impression of excitement and wonder."

"We were looking at things that human beings had never seen before," says Schmitt, "or if they'd seen them, they weren't thinking about them in terms of understanding our Earth and our solar system, and indeed the universe. And that's what we were, that's what we were doing. We were scientific explorers right from the moment that we stepped out of the spacecraft."

The twelve men who walked there would take thousands of pictures of lunar mountains, valleys, and rocks, but it was the photos of Earth that stood out for them as the real prize in the sky. "We'd spent all of this time with geology training and photographing the moon and understanding lunar craters; I mean we'd trained for everything... except for *Earthrise*," said Anders, "which really caught my attention and basically changed my life."

Anders noted further that it is impossible for anyone to comprehend from photos and words the scale of the hair-raising blackness of space. There is no atmosphere in the vacuum of space, which means there was nothing filtering the astronauts' line of sight and the surreal sensation of seeing, with superhuman-like vision, the only observable object of color in the solar system—our home planet. And because Earth has clouds, the sun's rays bounce off it like a colorful beacon in the surrounding void, turning it into a kind of cosmic lighthouse fifty times brighter than the moon, according to Lovell and the rest of the Eagles who've witnessed it.

This ocular mind trick—the laser-like interplay of light, darkness, and color between Earth, moon, and sun—stirred the artist in Bean, who devoted his entire post-NASA life to painting the missions to the lunar surface. On the day of our visit, he had studies of Earth in his art studio sitting side by side with his reproduction of Monet's *Water Lilies* to give him inspiration for his latest painting. The eighty-five-year-old Bean was transfixed with making Earth's colors as brilliant as Monet's masterpiece. This thoughtful teddy bear of a man, who named his Lhasa apso dogs Moon Beam and E.T., walked over to his easel with a fervent question of his own: "How can I make this image of the Earth more beautiful and still be somewhat representative of water lilies and flowers, although we've never, ever seen any water lilies and flowers that look that good...ever?"

Frustrated, he joked that if Monet were alive, he could give us the answer.

Looking at Bean's drafts of the *whole* Earth served as a powerful reminder of how nearly impossible it was for anyone looking at Anders's *Earthrise* photo, in 1968, to locate the seven continents. But even more disorienting was the jolting realization that one's home country—as an identifiable territory—had suddenly become a ludicrous concept.

Lovell took it one reductive step further. During Apollo 8's first of three television broadcasts, when they were 180,000 miles away from Earth, Lovell looked out his spacecraft's window and (thinking like an alien) posed this guileless, rhetorical question to the television audience. "What I keep imagining is if I am some lonely traveler from another planet what I would think about the earth at this altitude, whether I'd think it would be inhabited or not."

Not missing a beat, Collins, in Mission Control, shot back with a deadpan response worthy of a Stephen Colbert or David Letterman: "Don't see anybody waving; is that what you are saying?"

And there you had it. You couldn't see people, races, countries, or borders. Every reference point that man had ever organized himself around geographically, politically, and culturally vanished in the eyes of the Eagles and their Hasselblad cameras.

Was it even inhabited? This question especially struck Pulitzer Prize–winning poet Archibald MacLeish, whom the *New York Times* put on its front page, Christmas Day, to reflect on the impact of seeing Earth for the first time in the universe's rearview mirror:

> And seeing it so, one question came to the minds of those who looked at it (the astronauts). "Is it inhabited" they said to each other and laughed—and then they did not laugh. What came to their minds a hundred thousand miles and more into space—"half way to the moon" they put it—what came to their minds was the life on that little, lonely floating planet: that tiny raft in the enormous empty night. "Is it inhabited?"

MacLeish's peroration ended by articulating with exceptional grace what the greatest minds before him had hoped for—the sublime unity of mankind. It had been uttered by Buddha, Jesus, and all the great philosophers, but none of them had *Earthrise* in his hands to demonstrate *visually*, like a sound-and-light show, the truth of his message.

But the twentieth century did have it. For the modern mind, already steeped in photos, film, and television, the image of *Earthrise* fell on eyes primed to receive the possibility of the oneness of humanity. If a picture was worth a thousand words, MacLeish found the words worthy of the pictures of terra firma broadcast from Apollo 8:

> To see the earth as it truly is, small and blue and beautiful in that eternal silence where it floats, to see ourselves as riders on the earth together, brothers on that bright loveliness in the eternal cold—brothers who know now they are truly brothers.

"Riders on the earth together"! The Eagles, more than anyone, hoped for this earthly brotherhood to take hold—to become a movement. That nations would see what they saw from space and shed centuries of territorial bloodletting to join hands in new partnerships for peace and global cooperation.

When he addressed Congress to celebrate Apollo 8's historic voyage, Borman, the exacting West Pointer with no use for soft-headed thinking, read precisely the "Riders on the Earth" passage to its members and then expressed his hope that in a "few years we'd have an *international community* of exploration and research on the Moon."

Borman still shakes his head over the tepid applause he received for his proposal of brotherly cooperation.

Equally exasperated with politicians was Ed Mitchell, who, as the most mystically inclined of the Eagles, directed a solid punch at them for their inability to see beyond their limited local agendas:

> You develop an instant global consciousness, a people orientation, an intense dissatisfaction with the state of the world, and a compulsion to do something about it. From out there on the Moon, international politics look so petty. You want to grab a politician by the scruff of the neck and drag him a quarter of a million miles out and say, "Look at that, you son of a bitch."

"I really believe that if the political leaders of the world could see their planet from a distance of 100,000 miles their outlook could be fundamentally changed," writes Collins. "That all-important border would be invisible, that noisy argument silenced. The tiny globe would continue to turn, serenely ignoring its subdivisions, presenting a unified facade that would cry out for unified understanding, for homogeneous treatment. The Earth must become as it appears: blue and white, not capitalist or Communist; blue and white not rich or poor; blue and white, not envious or envied."

For the Eagles, the pettiness of international politics was a perceptual breakthrough made possible only by their deep space voyage. Mitchell, who died in 2016, summed it up this way: "We went to the Moon as technicians; we returned as humanitarians."

Political leaders may have been slow to process the implications of what the Eagles were saying, but in the larger culture it was a different story. Other than astronauts and movie stars, the only other people to achieve equal cultural status during the 1960s space age were rock stars. Rock stars were getting it! The biggest names in rock internalized not only the Eagles' images from space, but even NASA's lingo.

There was David Bowie's 1969 hit "Space Oddity," which has morphed over the decades into a kind of ageless anthem to otherworldliness, and "Space Cowboy" from the Steve Miller Band (also released that year), followed in 1970 by the Beatles' "Across the Universe," which hit a spiritual high note, and Elton John's elegiac classic "Rocket Man" from 1972, which climbed to number two on the charts in the UK, and number six in the US.

But it was environmental activists who recognized the immediate value of the Eagles' eyewitness testimony and photographs of Earth from space. More than anyone, they had cause to rally around *Earthrise*, which nature photographer Galen Rowell called "the most influential environmental photograph ever taken." On April 22, 1970, environmentalists launched Earth Day, marking the birth of the modern environmental movement, and an era of

"Earth consciousness" that we are still living in. Earth Day Network records that "20 million Americans took to the streets, parks, and auditoriums to demonstrate for a healthy, sustainable environment in massive coast-to-coast rallies. Thousands of colleges and universities organized protests against the deterioration of the environment."

CBS News anchor Walter Cronkite reported it this way:

A unique day in American history is ending, a day set aside for a nationwide outpouring of mankind seeking its own survival; Earth Day! A day dedicated to enlisting all the citizens in a bountiful country in a common cause of saving life from the deadly byproducts of that bounty—the fouled skies, the filthy waters, the littered Earth.

The unassailable weight of decades of visual, tactile, and olfactory evidence led Republican president Richard Nixon to create the Environmental Protection Agency, which opened its doors December 2, 1970, just seven months after Earth Day, to accomplish exactly what its name says.

Protecting the planet found its most holistic advocate in James Lovelock, a prominent British scientist (and NASA consultant), who took Earth consciousness right down to the molecular level. Lovelock, who refers to himself as a "planetary physician," named his hypothesis after the Greek goddess of Earth, Gaia. In his 1979 book, *Gaia*, Lovelock argues that Earth, when viewed from deep space, could finally be appreciated as a single, giant living organism in which all life forms are inextricably linked in a delicate, biospheric balance to preserve planetary health. "The living Earth complains of fever," Lovelock would write later. "I see the Earth's declining health as our most important concern, our very lives depending upon a healthy Earth."

Today the Earth's fever, as Lovelock described it, is called climate change. One of its more vocal apostles is former vice president Al Gore, who uses *Earthrise* and its sister image, the *Blue Marble*, taken by the crew of Apollo 17, as bookends to his two films documenting Earth's rising temperature: *An Inconvenient Truth* in 2006, and its 2017 sequel.

Acting on the accumulated scientific evidence of the last fifty years, by 2017, 195 nations had signed the Paris Agreement. Its primary goal is to reverse global warming by lowering carbon emissions from signatory countries. The United States was the only major power to withdraw from the Paris Agreement after President Donald J. Trump dismissed climate change as a "hoax."

Collins, in the meantime, worries that Earth is being mortally wounded. "The dead zone in the Gulf of Mexico is now larger than New Jersey and still growing. The growth of death: a terrible thing to do to this planet. This one example alone makes me want to cry and countless other catastrophes abound, some lurking in the future, many already here."

The awareness of the environment, generated by the *Earthrise* and *Blue Marble* photos, is evolving today in unexpected places—most intriguingly in the Divinity Schools at Harvard and Yale Universities, where a new field of study has emerged that is popular with young theology students, who worry about man's treatment of the planet. It is called Ecotheology. If Earth is God's creation, goes the argument, then Earth's life-giving environment needs to be nurtured with reverence and respect. "Because of the ecological crisis on the Earth," says the dean of Yale's Divinity School, Greg Sterling, "it is far more common now for theology not to focus solely on humanity, but rather to focus on the larger framework of Nature. Those photos of Earth allowed all of us to see our home in the universe from the outside and to realize THIS is what we have, and THIS is all we have for our own existence."

"The earth, our home, is beginning to look more and more like an immense pile of filth," mourns Pope Francis, whose 2015 encyclical letter, Laudato Si', is devoted entirely to climate change and mankind's moral obligation to protect God's creation "Mother Earth." Without calling it Ecotheology, the pope is actively leveraging the full weight of the church to build a theological case aligning religion with environmental goals. It has become one of the defining issues of his papacy:

> Never have we so hurt and mistreated our common home as we have in the last two hundred years. Yet we are called to be instruments of God our Father, so that our planet might be what he desired when he created it and correspond with his plan for peace, beauty and fullness.

Which brings us back to the Eagles. From out on the moon, sheathed in their liquid-cooled spacesuits and protective sun visors—a pinprick away from certain death if their armor failed— the Eagles came to a collective conclusion about where true paradise lay. The most unanticipated reaction, when he saw it, came from Al Shepard, who was known as the "Icy Commander." He was a *Mayflower* descendant who fought in World War II as a gunnery officer on a destroyer in the Pacific theater, shooting down kamikazes in the Battle of Okinawa. Shepard was renowned for a fearsome, almost pushy competitive toughness that was intimidating to more than a few of his equally formidable fighter jocks.

While Shepard worked away on the lunar surface, paradise had been hanging right above his gold visor. And when he finally looked up and saw its astonishing beauty, the "Icy Commander" blew his cool in a way that no one else had.

If somebody'd said before the flight... "Are you going to get carried away looking at the Earth from the Moon?" I would say, "No, no way!" But yet when I first looked back at the Earth, standing on the surface of the Moon, I cried.

With paradise in their sights, there was something else the Eagles would see, which in the words of Al Worden was "mind blowing"—the vastness of the universe. "The most wonderful thing I saw was the Earth. But the most awesome thing, and the thing that really got my imagination, was looking at the universe out there. That is so unbelievable, and then you begin to understand what it's all about, and where we are, and what we're part of, and that there's so much more out there. That's what's mind blowing."

And as the Eagles set course for home, it made them wonder: Is there a God? Could there be life on other planets? How did this all happen?

Chapter Six

THE EAGLES AND THE GOD QUESTION

The Mystic Mountain image taken by the Hubble
Space Telescope, 1997

KEY LESSONS

➤ Have faith that God and science
can coexist

➤ Connect not only to earth, but the
universe

➤ Find kindness and tolerance in all
that you do

➤ Accept doubt as an integral part
of faith

➤ Stand in awe of the unknowable and
embrace it

I find God is no fool. I think he probably spends a lot more time in the coral reef than he does in the cathedral.

Mike Collins, Apollo 11

Walking on the Moon was three days but walking with Jesus is forever. *Charlie Duke, tenth man on the moon*

Religion is man-made. The Creator I'm talking about stands above all those religions. *Gene Cernan, Apollo 17*

After the flight, you know, I thought this is ridiculous to think that God sits up there with his supercomputer," says Anders, who realized as soon as he got to the moon (and saw the Earthrise) that Catholicism's tenuous hold on him finally dissolved. "When I looked back and saw that tiny Earth, it snapped my world view," Anders said. "Here we are, on kind of a physically inconsequential planet, going around a not particularly significant star, going around a galaxy of billions of stars that's not a particularly significant galaxy—in a universe where there's billions and billions of galaxies...Are we really that special? I don't think so."

The lunar voyages of the Eagles would begin to add a new dimension to humanity's dialogue about the nature of God, metaphysics, and spirituality. The space age was only five years old when, in 1966, *Time* magazine's cover posed the question, "Is God Dead?" The great mythologist Joseph Campbell believed the moon landings upended every culture's mythological stories explaining God, heroism, and man's place in the universe:

"We are challenged both mystically and socially, because our ideas of the universe have been reordered by our experience in space," said Campbell, author of *The Power of Myth* and more than a dozen other books on mythology during his tenure at Sarah Lawrence College. "The consequence is that we can no longer hold onto the religious symbols that we formulated when we thought that the earth was the center of the universe.... With the moonwalk, the religious myth that sustained these notions could no longer be held. With our view of earthrise, we could see that the earth and the heavens were no longer divided but that the earth is in the heavens."

What Campbell described as the Earth being in the heavens, Ed Mitchell processed with his own eyes through the programmed rotation of his spacecraft, which gave him a view of Earth, the sun, the moon, and the heavens at two-minute intervals. Known as the barbecue roll, the rotation was required to sustain an even temperature throughout all parts of the spacecraft. That alternating view of each heavenly body, Mitchell reported, reordered his mind, his religious beliefs, and his own place in the universe in a radically different way from the rest of the Eagles. Mitchell had been raised a Southern Baptist in Artesia, New Mexico, and there was nothing in his background hinting of the transformation he was about to undergo.

He was a no-nonsense aircraft carrier pilot flying A3D Skywarrior fighter jets off the USS *Bon Homme Richard* and USS *Ticonderoga*. His next step up the right-stuff ladder was the US Air Force Aerospace Research Pilot School, where he graduated first in his class. He had an inquisitive, searching intellect, which led him to MIT, where he got his doctorate in aeronautics and astronautics. By then, Mitchell had already begun questioning his Christianity and was quietly exploring the fields of mental telepathy and paranormal phenomena, which he kept secret from his crewmates Al Shepard and Dick Gordon, who, like the rest of the Eagles, subsequently dismissed Mitchell as absurdly New Agey. "I always classify those feelings as loony tunes, and I still do," says Borman.

With the Earth, moon, and sun trading places outside his spacecraft's window, Mitchell was overwhelmed by an illuminating moment of clarity: "Suddenly I realized that the molecules of my body and the molecules of the spacecraft and the molecules in the body of my partners were prototyped and manufactured in some ancient generation of stars, and that was an overwhelming sense of *oneness*, of *connectedness* . . . I was overwhelmed with the sensation of physically and mentally extending out into the cosmos."

Mitchell preached his gospel of molecular mysticism through his Institute of Noetic Sciences, a nonprofit that is still dedicated to elevating human consciousness by focusing—through meditation—on man's connection to the universe.

When *Time* magazine asked the God question in 1966, 97 percent of Americans told Harris pollsters that they believed in God while only 27 percent considered themselves deeply religious.

Fast-forward to the Pew Research Center's 2017 poll, which found that a slim majority of Americans (56 percent) say they believe in God "as described in the Bible." One-third of Americans reported they do *not* believe in the God of the Bible, but nine in ten Americans believe in a higher power. Only one in ten do not believe in any higher power or spiritual force.

The Eagles fall all across this spectrum in their conception of God and spirituality, but the only two claiming to have experienced a revelation as a direct result of orbiting or walking on the moon were Mitchell and Jim Irwin, who was the eighth man to walk on the moon. In his autobiography, Irwin said God spoke to him on the lunar surface, giving him the solution to a difficult experiment he was trying to set up: "I had become a skeptic about getting guidance from God and I know I lost the feeling of nearness. On the Moon, the total picture of the power of God and His Son Jesus Christ became abundantly clear to me."

Irwin was born in 1930, in Pittsburgh, to Scottish-Irish parents who eventually moved to Utah, where Jim graduated from East High School in Salt Lake City. The calling to Christ came as early as age eleven, Irwin reported, when, by pure chance, he and his mother came upon a revival meeting at a Baptist church, where he decided amid all the shouts of *amen* that Jesus was his savior. But Jesus couldn't compete with the lure of flying. When he was twelve, Irwin announced to his mother that he would one day fly to the moon. His trajectory for getting there was typical of the Eagles ... first attending the US Naval Academy, where he received a BS in naval science, followed by an MS in aeronautical and instrumentation engineering from the University of Michigan. He was one of nineteen astronauts accepted into NASA's class of 1966 after completing stints at the Air Force Experimental Test Pilot School and the Aerospace Research Pilot School. Rediscovering Jesus on the moon led Irwin—post-NASA—to a devotional life through a foundation he called High Flight, named after a poem that was a favorite of many of the Eagles, including aviators today, who turn to it for inspiration. It was written by John Gillespie Magee Jr., an Anglo-American pilot who fought with Britain's RAF in World War II. He died in 1941, at age nineteen, during a training exercise in his Spitfire.

Oh! I have slipped the surly bonds of Earth
And danced the skies on laughter-silvered wings.
Sunward I've climbed, and joined the tumbling mirth
Of sun-split clouds—and done a hundred things
You have not dreamed of—wheeled and soared and swung.
High in the sunlit silence, hov'ring there
I've chased the shouting winds along and flung
My eager craft through footless halls of air.
Up, up the long delirious burning blue
I've topped the wind-swept heights with easy grace
Where never lark nor ever eagle flew
And, while with silent, lifting mind I've trod
The high untrespassed sanctity of space,
Put out my hand and touched the face of God.

Irwin would be the first and youngest of the moon walkers to die at sixty-one from heart failure in 1991. Like Mitchell, many of the Eagles stopped taking him seriously when he began leading expeditions to find Noah's lost ark at Mount Ararat, in Turkey, where he concluded that the Bible's seven-day creation story was God's truth.

While they were still working at NASA, the Eagles, however, were careful not to trumpet their religious beliefs (or lack of them) since they had no interest in tipping the scales, either way, in the highly charged debate over God's existence. But in the spirit of America's heritage of religious freedom, neither did NASA want to prevent anyone in the agency from invoking the name of God. The most memorable instance came in 1962 as John Glenn cleared the launch tower in his Mercury capsule, Freedom 7. It would go down as the most quoted launch send-off in space history: "Godspeed, John Glenn," his colleague Scott Carpenter called out as the nation prayed for Glenn to become the first American to successfully orbit Earth. Nor did NASA want to script anything the Eagles would say to the public during their live TV broadcasts from space... including Neil Armstrong's eternal first words from the surface of the moon: "That's one small step for man, one giant leap for mankind." NASA's leadership trusted the Eagles' judgment, and at no time was this more evident than on the flight of Apollo 8, when NASA's public affairs director, Julian Scheer, told Frank Borman to simply "say something appropriate" for their scheduled Christmas Eve broadcast from the moon.

This bedrock American principle of "freedom of speech" would be in stark contrast to the closed Soviet space program, where cosmonauts hewed to the Communist Party line, which outlawed religion or any mention of God. It was widely reported, for example, that Yuri Gagarin, the first human to orbit Earth, said during his flight: "I looked and looked but I didn't see God." Historians now believe that it was probably Soviet premier Nikita Khrushchev who fabricated Gagarin's quote to prove that God was a capitalist concoction duping the masses into believing that a better life awaited them as they were being exploited to death. It was a direct swipe at the West and America, whose presidents took their oath on the Bible and routinely cited God in their speeches, just as President Kennedy did for the lunar challenge... "And, therefore, as we set sail we ask God's blessing on the most hazardous and dangerous and greatest adventure on which man has ever embarked."

Borman, a committed Protestant, naturally laughed off Gagarin's clumsy quote, saying, "I didn't see God on my first spaceflight either, but I had an enormous feeling that there had to be a power greater than any of us—that there was a God, that there was indeed a beginning."

As for the Christmas Eve lunar broadcast, Borman believed the event required a note of planetary grace resonant for all humanity. Lovell and Anders also appreciated that this would be no ordinary broadcast. It would mark our solar system's first interplanetary transmission emanating from men circling the moon in a direct address to Earthlings. It had the potential for literary majesty. What words, what statement for the people on Earth could possibly match the historical magnitude of the moment without sounding provincial and trite? Since wordsmithery wasn't his strong suit, Borman asked Scheer for help. "I think it would be inappropriate," replied Scheer, "for NASA and particularly for a public affairs person to be putting words in your mouth. NASA will not tell you what to say." Borman, Anders, and Lovell conferred among themselves about what qualified as appropriate. Perhaps something related to Christmas, they thought, but that boxed them in to the birth of Jesus, which would have been fine for the 1.1 billion Christians on the planet, but not for the Muslims, Jews, Hindus, Buddhists, and other faith groups who might feel offended or left out. The Eagles were already starting to think on a planetary scale, says Borman. Their earlier Gemini flights orbiting the Earth had already planted the seed that Earth, viewed from space, was a seamless mass without visible borders, nations, or peoples. They wanted something to reflect this new awareness but couldn't come up with the words to express it.

Borman decided to turn to his old friend Simon Bourgin, who worked at the United States Information Agency and was a former newspaperman. "He had a worldly sophistication," says Borman, "but also knew how to write for a popular audience." Borman made it clear to Bourgin that whatever statement he penned needed to be inclusive. But Bourgin soon hit the wall, too, unable to find anything that would have meaning across multiple cultures. He in turn handed the assignment to Joe Laitin, a colleague who had spent ten years in Hollywood as a television and magazine writer. Laitin started his career as a World War II correspondent and wound up as an assistant press secretary for President Johnson. Laitin knew how to write dramatic moments and thought he might find something inspirational and poetic from the New Testament. "You're in the wrong part of the Bible," his French wife, Christine, told him as he was thumping away on his typewriter at four in the morning. "Everything I wrote," said Laitin, "was about peace on earth and here we were in a war in Vietnam! Writing about peace on earth at that time would have made us all look stupid. By now, the floor was littered with balls of paper, and I was beginning to get a little frustrated, because I considered myself a professional." Christine was a prima ballerina in Paris at the start of the German occupation, an intellectual who read a book a day. She pointed her husband straight to the beginning: Genesis. "In the beginning God created the heaven and the earth," ran the first line of the first verse. Laitin knew right away, as did Borman, Anders, and Lovell when they saw it, that he had found the perfect statement. The overarching story of God creating heaven and earth was common to enough cultures that it felt exceptionally suited to man's first look at Earth from another heavenly body.

At 9:31 p.m. Eastern Standard Time, on December 24, 1968, Anders opened with what would become known as the Genesis broadcast, marking one of the most spiritual moments in television history:

We are now approaching the lunar sunrise, and for all the people back on earth, the crew of Apollo 8 has a message that we would like to send to you:

In the beginning, God created the Heaven and the Earth. And the Earth was without form and void, and darkness was upon the face of the deep. And the Spirit of God moved upon the face of the waters. And God said, "Let there be light." And there was light. And God saw the light and that it was good, and God divided the light from the darkness.

Lovell and Borman would each read the next nine verses, with Borman closing the broadcast, saying: "And from the crew of Apollo 8, we close with good night, good luck, a Merry Christmas—and God bless all of you, all of you on the good Earth."

Back in Mission Control no one had been told that the crew would read from the book of Genesis. For those in the room, it generated a near parting-of-the-waters moment that left many in tears as they absorbed what they had just heard. "It was a religious experience to me," recalled John Aaron. "When they started reading from the book of Genesis, that was like, 'Wow!' The hair on the back of my head, on the back of my neck, actually stood up. I had that kind of feeling."

"I was enraptured, transported by the crew's voices, finding new meaning in the words from Genesis," writes Gene Kranz. "For those moments, I felt the presence of creation and the Creator. Tears were on my cheeks."

"A scriptwriter couldn't have done a better job," Lovell says today. "Reading from the first ten verses of Genesis was the capstone of our last orbit, and it brought people back on Earth a little closer together."

Not quite everyone interpreted the broadcast that way. It would, of course, have been almost un-American if someone in the country didn't find legal fault with the Genesis broadcast, and it was left to Madalyn Murray O'Hair to deliver on America's proclivity for litigiousness. O'Hair was the founder of American Atheists, who in 1963 won the Supreme Court case ending Bible reading in American public schools. O'Hair filed suit against NASA, claiming it had violated the First Amendment by allowing the Apollo 8 crew to make what she considered a religious statement from deep space. She lost her case but from then on, the Eagles became a bit more muted in expressing their religious views while in space.

As the years went by, Anders revealed that while the Genesis broadcast set the perfect tone for their message to Earthlings, it carried no spiritual significance for him. Anders says he was more concerned about not disappointing Borman, a lay deacon whose faith is an integral part of his being. Today they kid each other about the differences the moon voyage had on their faith. "He's a wonderful friend," says Borman, "but I think that the experience of Apollo 8 led him to believe that there was so much more than just what we humans are, and he wondered how could 'One Power' be the ultimate."

Borman came to the opposite conclusion as he stared at Earth (and the surrounding vastness of the universe) thinking, "This must be what God sees." And even though he read from Genesis on Christmas Eve, he is hardly a creationist, as he explained to a Lutheran minister whose church he occasionally prayed at because it's close to Susan's nursing home. Borman says the minister unceremoniously grilled him about his beliefs because he was not a member of his church and was hesitant to let him attend:

"And he asked me, 'Do you think God created the Earth?'

"And I said, 'There's no doubt in my mind he created the Earth, and the universe too!'

"And he said, 'Well—do you think he did it as the Bible says in seven days?'

"And I said, 'No! I'm sure he didn't, I've studied geology.'

"'Well, what do you think about women pastors?'

"And I said, 'I don't have any trouble with them as long as they meet the same requirements as the men do.'

"And then he asked, 'Well, what do you think about homosexuals? Do you think homosexuality is a sin?'

"I said, 'No! I don't think it's a sin, I think people are born that way,' and I said, 'You have to be tolerant.'"

At the end of this blunt interrogation, Frank says the minister told him his answers were wrong on every question. "So, to me," Borman concluded, "that was the most unchristian church in the world," and he never went back.

After Apollo 8 there would be no more religious readings coming from the Eagles. When Armstrong and Aldrin made their landing, Aldrin, a Presbyterian, would offer only a very generic statement of gratitude for their safe arrival: "I would like to request a few moments of silence . . . and to invite each person listening in, wherever and whomever they may be, to pause for a moment and contemplate the events of the past few hours, and to give thanks in his or her own way." And then, privately in the presence of Armstrong, who was a deist, Aldrin took communion before stepping onto the lunar surface: "I poured the wine into the chalice our church had given me. In the one-sixth gravity of the Moon the wine curled slowly and gracefully up the side of the cup. It was interesting to think that the very first liquid ever poured on the Moon, and the first food eaten there, were communion elements."

Armstrong, on the other hand, would remain an enigma regarding religion. The Armstrongs were Presbyterians, and his devout mother, according to Neil's authorized biographer, James R. Hansen, never imposed her faith on her children. Atheists, like O'Hair, would try to claim Neil as one of their own, but they had no proof, and no interviewers ever got close enough to buttonhole him with the direct question "Do you believe in God?" The closest anyone came, says Hansen, was Walter Cronkite, on CBS's *Face the Nation*, when he interviewed Neil three weeks after he returned from the moon. Cronkite's question pertained to O'Hair's statement that Neil was, in fact, an atheist. "I don't really know what that has to do with your ability as a test pilot and an astronaut, but since the matter is up, would you like to answer that statement?" Neil replied, "I don't know where Mrs. O'Hair gets her information, but she certainly didn't bother to inquire from me nor apparently the agency, but I am certainly not an atheist." Cronkite didn't stop there: "Apparently your [NASA astronaut] application just simply says 'no religious preference.'" Hansen writes that "as always, Neil registered another answer as honest as it was vague and nondescript." "That's agency nomenclature," said Neil, "which means that you don't have an acknowledged identification or association with a particular church group at the time. I did not at that time."

And with that Cronkite dropped the entire line of inquiry. There was one written record, however, that did capture Neil's religious affiliation when he applied to lead a Boy Scout troop at a Methodist church, in Southern California, in the late 1950s. The application asked him to list his religion...to which he replied, "*Deist.*" Hansen writes, "The confession so perplexed the Methodist Minister that he consulted Stanley Butchart, one of Neil's fellow test pilots as well as a member of the congregation. Though uncertain of the principles of deism, Butchart praised Neil as a man of impeccable character."

Deism was a particularly interesting choice for Neil to make. It's how Thomas Jefferson thought of himself. Jefferson spent the last few years of his life writing the Jefferson Bible, in which he redacted the miracles and incongruous statements attributed to Jesus (by later Bible writers) whose words Jefferson believed did not match the cadence and revelatory genius of Jesus's beatitudes, also known as the Sermon on the Mount. The best understanding of Jefferson's, and today's, deism comes from the *Cambridge Dictionary*, which describes it as "the belief in a single God who does not act to influence events and whose existence has no connection with religions, religious buildings or religious books."

Other than the Boy Scout application, Armstrong, in keeping with his intense desire for privacy, would never speak of deism or make any public comments about religion for the rest of his life. No other Eagles considered themselves deists. Some, like Anders and Collins, came closer to Albert Einstein's agnostic contemplations of religiosity, in which he plainly stated that he was not an atheist but was in reverential awe of the unseen hand ordering the universe through the laws of physics.

For Collins, the God question remains a vexing mystery since he is neither an atheist nor an unquestioning believer. "I'm not an atheist. I think that the people who deny the existence of God," Collins says, "are as arrogant as some of the born-agains, who've got the answer to each and all problems." Collins, like Anders and many of the Eagles, is drawn to the kind of enlightened agnosticism articulated so compellingly by Einstein. Einstein points to the observable order of the universe as the source of religiousness and the possibility of an intelligent creator.

The most beautiful and most profound emotion we can experience is the sensation of the mystical. It is the sower of all true science. He to whom this emotion is a stranger, who can no longer wonder and stand in rapt awe, is as good as dead. To know that what is impenetrable to us really exists,

manifesting itself as the highest wisdom and the most radiant beauty which our dull faculties can comprehend only
in their most primitive forms—this knowledge, this feeling
is at the true center of religiousness.

"I'm a big believer in Einstein," says Collins. "The idea that
he didn't need any computers; he didn't need the telescopes. All
he needed was chalk and a chalkboard and with those two things,
those were his weapons. He could figure out things that no one else
can come close to figuring out. If you took the smartest little baby
and trained that little baby to be an astrophysicist or whatever, they
wouldn't come close to being as intuitive."

Collins had been raised in a Catholic family, but his mother's church attendance, he believes, was more for social purposes
than religious. "My father," says Collins, "dropped out of the faith
because he told me that the nuns just lied to him. He never would
expand on that other than to just say that." Collins attended the
prestigious St. Albans School, an all-boys Protestant enclave for
the nation's ruling class, which includes such prominent alumni
as former vice president Al Gore. It was at St. Albans when Collins remembers having his first misgivings about religious truth. It
happened, he says, at about age sixteen when he realized that there
were dozens of religions in the world and that not all of them could
claim to be God's chosen faith. "That was shocking," said Collins,
"to realize if the Episcopalian faith is right, then all the others must
be wrong. And I asked my mentors at the school to explain it, and

I got a bunch of blather that I did not understand then, and I don't understand to this day."

Where Collins places his faith today is in the big bang theory, which says the universe originated billions of years ago in an explosion from a single point of nearly infinite energy density. "If we start with the big bang," says Collins, "somebody, somewhere, something, some force, person, he, she...to me that's God." And Collins further notes that he finds God to be more present in nature than in church. "So, when I go to a church service"—and here Collins is unsparing—"and that old fat guy comes waddling down the aisle waving his incense pot, my brain flies right out of the window, beautiful though the cathedral is, and goes to a coral reef. I find God is no fool. I think he probably spends a lot more time in the coral reef than he does in the cathedral."

Gene Cernan, who died in 2017, was another Catholic who, after his walk on the moon, found the church's teachings provincial and earthbound. Flying around a sliver of our solar system had a way of doing that to some of the Eagles. It was one thing for Einstein to put his theories on paper and chalkboard, as Collins observed, but something viscerally different to see his formulas (along with the Newtonian laws of gravity) proven in deep space. The lives of the Eagles depended on those universal laws of physics. Theory turned into exhilarating reality as the Eagles entered the moon's gravitational pull at the precise moment NASA's engineers said they would. That the moon's one-sixth gravity was exactly as predicted. That they would lose and reacquire radio contact with Houston, just as the flight plan projected they would after entering the moon's far side. It was that precise. It was that orderly. There were no surprises. Their lives hung in the balance of those mathematical formulas. And when experienced 230,000 miles from home, centuries of theological doctrine suddenly became quaint in the backdrop of billions of luminous stars ordered with such precision. For Cernan the solar system had become his new church. The godly symmetry of it all led Cernan to revise his view of Christianity: "Religion is man-made. The Creator I'm talking about stands above all those religions," Cernan said. "The Earth doesn't tumble through space; it moves with logic and certainty and with beauty beyond comprehension. It's just too beautiful to have happened by accident. Someone—some being, some power—placed our little world, our sun and our moon where they are in the dark void."

Alan Bean said that he, too, like Einstein, was in "rapt awe" of his celestial surroundings, but that his gradual separation from God had nothing to do with his thirty-one hours spent traversing the moon's Ocean of Storms. Bean was a Methodist who taught Sunday school in 1960 while he was a test pilot at Patuxent River, Maryland. When Bean had his children, Clay and Amy, he began rethinking the whole notion of "God the Father." A father's role (God's role), Bean reasoned, was to protect his children, but where it all broke down for him was God allowing dreadful things to happen to innocent people: "Just go to a cancer ward somewhere," said Bean. "I wouldn't treat my kids that way. My opinion is, if there is a God, he's doing a shitty job taking care of us, and as I look around, he's not so good to some people."

Bean said he couldn't reconcile life's inexplicable cruelties and tragedies with a supposedly fatherly God. He thus stopped believing. Bean was quick to note, however, his gratitude for his own worldly success and for being spared suffering in his life. "And that's why I joke sometimes if there is a God, he's probably looking down at me and saying, 'You asshole, look what I've done for you.'"

Although he was not a practitioner, Bean said, later in life, he found greater truth in Buddhism, which has no God but celebrates and puts nature and kindness at the center of its belief system. "I like to treat everyone with loving-kindness," Bean said. He also agreed, as did all the Eagles (save Irwin), with Enrico Fermi's statistical analysis that life likely exists elsewhere in the universe given the presence of so many billions of galaxies with stars similar to our sun that might have Earthlike planets. Fermi was an Italian nuclear physicist who helped America build its first nuclear bomb as part of the Manhattan Project. What ultimately perplexed Fermi is what came to be known as Fermi's paradox, in which he famously asked, "Where is everybody?" Why, he wondered, if there are billions of planets older than Earth, have advanced civilizations not come forth to contact us? The only Eagle who believed this had already happened was Mitchell. While Mitchell never claimed to have seen UFOs or extraterrestrials, either in space or on Earth, he believed in their existence—including the famous Roswell Incident, in which an alien spacecraft reportedly landed in Roswell, New Mexico, and was covered up in an ongoing conspiracy by the US Department of Defense.

What all the Eagles agree on is that, when viewed from the deadness of the moon, life on Earth is a miracle, a living paradise that much of humanity has failed to respect and care for. And if life does exist on other planets, it is not incongruous with the faith of believers like Borman, who thinks everything in the universe is part of God's cosmic tapestry... "And none of us will really know," he believes, "until we finally die." Collins, meanwhile, ponders the space-age version of the "Who created the Creator" question. "I ask myself, what was there before the big bang?"

Chapter Seven

PARADISE FOUND: AND ON THE BRINK

In 2017 NASA's Lunar Reconnaissance Orbiter captured a
stunning view of Earth from the spacecraft's vantage
point in orbit around the moon

KEY LESSONS

➤ Take care of the only home we
have: it is the crown jewel of the
universe

➤ Be mindful of earth's interconnectedness; our tenure is fragile

➤ Look into the night sky and ponder
the mystery of things

Trust me, there is no place in the universe like it.   *Alan Bean*

We need a new economic paradigm to produce prosperity without growth.
                                   Mike Collins, Apollo 11

Science and fairy tales are hardly strangers to each other, and there is one classic in particular that astronomers and astrobiologists have borrowed from to explain Earth's charmed place in our solar system. They call it the Goldilocks Zone.

The Earth sits at an ideal distance from the sun, the Goldilocks Zone (a life-creating sweet spot), making it neither too hot nor too cold for life to flourish. Mars, by comparison, is too far from the sun, and thus too cold for the ingredients of life to have evolved; Venus, being too close to the sun, is like a boiling sauna, making life equally impossible. Of the dozens of robotic probes that have explored our solar system in the past six decades, not one has spotted a planet containing the right mix of water, oxygen, and habitable temperature.

Alan Bean was awed by Earth's singularly matchless characteristics: "My feeling is, in comparing it to any place else we've ever seen, either with a telescope or Voyager or anything else—we are living in paradise. We're the only place that looks like this, the only place that has water where life-forms could evolve. Without water, without some medium where simple molecules can get together, you can't ever get complicated molecules like us. So as far as we know, *this is it*—this little spot in the universe is the only place that we've seen that you could evolve human beings, or even a rabbit, or even a tree. We've never seen any of that but here."

"From low earth orbit," says Mike Collins, "you see the good of the ocean, the good of the land, the good of the beautiful cloud formation, the sunlight. It's a dynamic, alive place, more awe inspiring and more beautiful than you imagined. Not so at the moon, where it is utterly barren, no motion."

"Apollo 8 is when I first looked at the Earth, as it really is, in its proper perspective to our solar system," says Jim Lovell. "We're the only planet that has the conditions that foster life as we know it." Lovell says his reading list these days includes a lot of books on cosmology, and he's intrigued by the discovery of exoplanets, planets that exist outside the solar system with the possibility of extra-terrestrial life: "Our galaxy," Lovell notes, "has millions of stars, and our galaxy is only one of millions of galaxies. And science has so far found hundreds of exoplanets, bodies that are going around stars." And some of those, he speculates, "have to be in the life position"—the Goldilocks Zone—"at the right distance to contain water, so we'd be very naïve to think that we're the only intelligent life in the universe."

Water is the operative word in so many of the descriptions of Earth given by the Eagles. For those of us who are earthbound, it is easy to forget that 71 percent of the planet is *water*. There is no vantage point on Earth, no mountain or perch from which to get any sense of the scale of the seas and oceans versus the rest of the planet or their role in sustaining it. Water is the lifeblood of the Earth . . . as it is of the human body, which is 60 percent water. Without food the average person can survive for approximately three weeks, without water only four to seven days.

It is only from space that one comprehends the extent to which water stands out as the most striking feature of the planet, separating it from everything else in the solar system. It is one of the first things Buzz Aldrin described: "Standing on the Moon looking back at Earth . . . you see all the colors, and you know what they represent. Having left the water planet, with all that water brings to the Earth in terms of color and abundant life, the absence of water and atmosphere on the desolate surface of the Moon gives rise to a stark contrast."

To see the exquisiteness of the Earth in contrast to the surface of the moon was quite striking. It was a cosmic aesthetic whose power caused Al Shepard to cry. For Lovell it gave him a new set of eyes and awareness of our home planet's grandeur. For fifty years, Lovell says, he's been reflecting on what we have "back here on Earth" and the improbability of so many rich life-forms and lush vistas finding their way to our celestial doorstep. "A beautiful sunrise, the feel of a fresh breeze, and rainstorms" are what profoundly move Lovell today, much like Collins finding God in a coral reef. "You walk here to this park, take a look at the lake, and the grass that's growing and the trees that are coming up. *This is it!* You're living! This is the best time you're going to have."

And that is what Lovell says led to his conclusion that "we

don't go to heaven when we die, we go to heaven when we're *born*." It's a striking transposition of the New Testament's "ascension" narrative (rising to heaven *after* death) and the great stories of mythology, which have looked beyond Earth to the heavens as eternity's terminus.

"It's a powerful formulation, and I don't know anyone who has articulated Lovell's sentiment in those words," says Yale's Greg Sterling. "It's born of that experience he had seeing Earthrise from space. There was a sense he had of already being in heaven in his life in a way that other human beings were not, and that perspective from space gave him a different understanding." "There is a heaven," says Lovell, "and God has given you the opportunity to enjoy the best things in life right here on Earth, right now."

"Brilliant, genius, it's great," says Bezos when he first hears Lovell's comment, "because this is the best planet. That's what he's talking about. You're here."

The Eagles, as test pilots, already had a built-in sensitivity for absorbing everything in the *present moment* and what was in front of them, but their lunar flights gave them something extra and unanticipated. It elevated their observational powers into a hyperawareness of Earth's elemental qualities. Take something as basic as the air we breathe. In space there is no breathable air, since it's a vacuum. The Apollo spacecraft carried 100 percent oxygen at a pressure of five pounds per square inch, which recycled the astronauts' carbon dioxide through the lithium hydroxide canisters that became so critical on Apollo 13. For the average eleven-day round-trip to the moon, the Eagles felt no wind, no atmosphere, no natural oxygen.

Worst of all was the wretched stench produced by their toilet procedures, which in zero gravity made it difficult for urine and bowel movements to remain fully captured in the hoses and plastic bags they used to get the job done. When they finally splashed down in the Pacific Ocean and opened the hatch, it was like "landing in the cradle of life," said Bean. The rush of the fresh, moist Pacific air, he said, was akin to a resurrection after being sealed in an artificially controlled environment for so many days without seeing a shred of greenery or any life-forms. Henceforth, Bean said, he felt nothing but gratitude for even the most mundane earthly happenings we all either ignore or are annoyed by: "I have not complained about the weather one single time. I'm glad there is weather. I've not complained about traffic. I'm glad there are people around. One of the things that I did when I got home, I went down to the shopping centers and I'd just go around there, get an ice cream cone or something, and just watch the people go by and think, 'Boy we're lucky to be here. Why do people complain about the Earth?' We are living in the *Garden of Eden*."

"I can remember the beautiful water," says Collins of his splashdown. "And we're out in the deep ocean of the Pacific and it was such a startling violet color, and I remember looking at the ocean and admiring it—'nice ocean you've got here, planet Earth.'"

Upon returning to their home planet, Neil Armstrong and the Eagles would start visualizing the Earth in a radically new way. From their vantage point in deep space, Earth was a spaceship whose inhabitants were living on the outside as they circled the sun.

"Hopefully, by getting a little farther away, both in the real sense and the figurative sense," said Armstrong, "we'll be able to make some people step back and reconsider their mission in the universe, to think of themselves as a group of people who constitute the crew of a spaceship going through the universe. If you're going to run a spaceship, you've got to be pretty cautious about how you use your resources, how you use your crew, and how you treat your spacecraft."

Political leaders, like Democrat Adlai Stevenson and his friend British economist Barbara Ward, had already begun sounding Earth's fire alarm using the same spaceship analogies, but with blunter warnings. Here's what Stevenson, who twice lost the presidency to Dwight D. Eisenhower, told the United Nations in 1965:

> We travel together, passengers on a little space ship, dependent on its vulnerable reserves of air and soil; all committed for our safety to its security and peace; preserved from annihilation only by the care, the work, and, I will say, the love we give our fragile craft. We cannot maintain it half fortunate, half miserable, half confident, half despairing, half slave—to the ancient enemies of man—half free in

a liberation of resources undreamed of until this day. No craft, no crew can travel safely with such vast contradictions. On their resolution depends the survival of us all.

To Stevenson and other influential thinkers of the 1960s, the planet was perforated with dichotomies of famine and abundance, dividing the developed countries from those desperate for resources. By their estimation, Earth was certainly no paradise, and for most of its inhabitants daily life was a physical and mental minefield. Their suffering was further intensified by the increasingly abusive treatment of the planet by marauding industrial polluters. Picking up on Stevenson's planetary theme, Ward decided in 1966 to document Earth's glaring environmental and social inequities in a book aptly named *Spaceship Earth*. She is credited with helping lay the foundation for the "sustainable movement" and became an apostle for balancing humanity's needs without straining the Earth's finite resources: "The careful husbandry of the Earth is *sine qua non* for the survival of the human species," she wrote, "and for the creation of decent ways of life for all the people of the world."

Ward, above all, argued for the right of every human being to an adequate standard of living, a plea that struck right at the heart of America's inner cities, which were engulfed in social and economic chaos. If anyone needed a moon shot, argued civil rights activists, it was the people who couldn't put food on the table. Their leaders were openly hostile to the space program, which they saw as a vainglorious pursuit shortchanging the poor at every turn. Jesse Jackson was chief among them, telling the *New York Times* on July 21, 1969, right after Armstrong landed on the moon:

> While we send men to the Moon, or send deadly missiles to Moscow, or toward Mao, we can't get foodstuffs across town to starving folks in the teeming Ghettos. While our astrophysicists can figure out the formulas that make the amazing trajectories and landings possible, we can't seem to get nutritionists and physicians to the shanties and shacks of Appalachia. Even as astronauts stride forth in the heady-ing atmosphere of the moon blindfolded, America moves toward the whirlwind of another long, fiery summer and on to more campus rebellions and bloodletting come Septem-ber. Thus, I bid us temper our shouts of exaltation as man breaks the fetters of gravity while being unable to forge the links of brotherhood.

The disconnect between the space program and the rest of American society was in plain view. On one side of the ledger, life-changing technological gains were made—mostly underappreciated at the time. On the other side, political and social disintegration roiled the country with disturbing regularity. Robert F. Kennedy and Martin Luther King were both assassinated in 1968, and the Vietnam War's divisiveness led to deadly confrontations in the streets. Four students were shot dead at Kent State University in 1970 during campus protests that would continue at other colleges for another three years. At Ivy League schools the space program had little or no support, and in some cases there was outright hostility. Frank Borman remembers being booed at Columbia University, then a center of anti-war and anti-establishment ferment, where students started throwing marshmallows at him during his college speaking tour about Apollo 8. At Cornell he faced similar treatment and unsympathetic questions over the cost of the program: "How can you spend all this money going to the Moon when there are so many poor, so many economic inequities, so much poverty?" the students wanted to know. Borman says he tried to explain that the "space program was a natural extension of the human mind, that it was something our society could not afford to abandon because the future—the students' future—depended on our keeping up with advanced technology."

After Apollo 9, Dave Scott, Rusty Schweikart, and James McDivitt found themselves making the same proactive case at a private White House dinner with President Nixon, which included an outspoken Democratic senator who also questioned the program's value. Armstrong had yet to land on the moon, and the senator doubted it would ever happen: "We've wasted all this money," the unnamed senator protested. "What do you think you're really going to get out of it?" he asked Scott. "Why spend money on exploration when you can spend money on the homeless?" "We talked about the benefits for science and technology, exploration, spin-offs for the American economy," writes Scott. "We argued then, as people continue to argue now, that investing in space exploration has to be a conscious decision by society to invest in the future. If you don't spend money on space," Scott continued, "it doesn't automatically follow that money will be spent on the needy and homeless."

Humanity's yearning to explore, versus its earthlier obligations, is one of those hamster wheel debates that leaves each side immutably passionate about its rightness. Jeff Bezos, for example, is so convinced of man's evolutionary imperative in space that he is spending $1 billion per year of his personal fortune for interplanetary exploration and the new discoveries that will be made as a result. The scale of his effort has predictably resurrected the same questions faced by Borman and Scott fifty years ago. As recently as 2018, a public policy expert at the University of Chicago, Dr. Harold Pollack, wrote a *New York Times* editorial asking Bezos to halt his space venture in favor of needy students, prevention of communicable diseases, and other worthy causes.

Bezos's and NASA's position remain constant: that space exploration is the Rosetta stone to a deeper understanding, not just of the universe, but of our own home planet. They point to the space program as more than just a job creator but as an incubator for brilliant innovations that can help save the home planet in unforeseen ways (as weather satellites proved through their preemptive ability to save countless lives from approaching killer storms, and other natural catastrophes).

Wherever one comes out on the space-versus-Earth spending debate, there is no argument over the Goldilocks exceptionalism of *Spaceship Earth*. Even from low earth orbit it continues to beguile and absorb each new generation of astronauts who've looked at it from either the retired space shuttle, or the International Space Station (ISS). "The Earth just looked alive and glowing the first time I looked at it," says recently retired astronaut Nicole Stott, who spent 104 days living and working on the shuttle and ISS. "We tend to think of ourselves in an isolated way" says Stott, "my neighborhood, my house, my state, or whatever it might be. When in fact everything absolutely is interconnected, and when you pull away from Earth, the way the Apollo guys did, you see it as *one* home."

Our home planet's interconnectedness, described by today's astronauts, is exactly James Lovelock's Gaia hypothesis, which has now been formally adopted and accepted as Earth System Science (ESS). ESS is defined as "the application of systems science to the Earth sciences. In particular, it considers interactions between the Earth's 'spheres'—atmosphere, hydrosphere, cryosphere, geosphere, pedosphere, biosphere and, even, the magnetosphere as well as the impact of human societies on these components."

Collins says he believes in Gaia/ESS now more than ever. "I didn't think of her at the time, but I've thought of Gaia a lot since then. Gaia was kind of in charge of what I saw. I summon it up when I'm writing, or I'm having a conversation about the Earth, its fragility, where we're going, and what we human beings are doing to it. I pretend to converse with Goddess Gaia and it's helpful to me in that regard."

Collins spent part of his post-NASA career as the director of the National Air and Space Museum, followed by work in private industry. Today, he spends most of his time fishing, painting, and attending lectures on the space program, to which he's devoted more ink than any other astronaut—four books in all—with *Carrying the Fire* still considered one of the finest in space literature. One of Collins's more forward-looking and thought-provoking statements has been that "we need a new economic paradigm to produce prosperity without growth." Unfettered business growth, he argues, has been ruining the planet, something that many environmentalists have been saying for decades.

"The economic paradigms are completely divergent between economists and ecologists," says Phil Duffy, president and executive director of the Woods Hole Research Center in Massachusetts, which is considered one of the leading think tanks on climate change. "Any scientist knows that 3 percent growth, along with the current levels of resource extraction, cannot continue ad infinitum and the challenge of creating 'prosperity without growth' as Collins stated, hasn't even begun to enter the consciousness of the business community."

Stewart Brand, who in 1968 created one of the early bibles of the ecology movement, says the stakes have gone way up since he first published the *Whole Earth Catalog*, which used *Earthrise* as its cover photo and, among other things, taught people how to live off the land. Brand was a disciple of Buckminster Fuller, a futurist-environmentalist who built the first large-scale geodesic dome in Montreal in 1967 and believed in "doing more with less." "Climate change," says the eighty-year-old Brand, "is still the most serious thing going on that humanity is facing in this century, and the scale of forces this time is planetary."

"Earth has changed a lot since we started flying in Gemini," said John Young, who died in 2018. "There's a lot of things like urban pollution and you can see that when you hit orbit now. You can see the big cities all have their own set of unique atmospheres, they really do. We oughta be looking out for our kids and our grand kids, and what are we worried about? The price of a gallon of gasoline. Here in the United States we're worried about three dollar a gallon gas. That's awful."

"It truly is an oasis," says Scott, "and we don't take very good care of it, and I think the elevation of that awareness is a real contribution to saving the Earth."

"I do worry about climate change," agrees Frank Borman, "and we have to find ways of reversing it."

Anders, who got his master's in nuclear engineering, takes climate change seriously enough that his foundation donates to educational and environmental causes. "Solving it," he says, "requires international cooperation," and like Brand and Lovelock he believes that nuclear power is the fastest and cleanest way to lower carbon emissions and meet the planet's energy demands. But he argues that nuclear power is currently not a politically popular solution and when Lovelock, who is ninety-nine years old, made the case for it in the early 2000s (and still does) he enraged his fellow Green and environmentalist friends. For them nuclear power is always one step away from calamities like Chernobyl and the Fukushima reactor in

Japan, which suffered a meltdown after being overwhelmed by the one-two punch of an earthquake followed by a tsunami in 2011.

Sifting through the current climate change debate and the subtly shifting positions of its various stakeholders is a complex affair, and it's reflected in the way the Eagles' views have evolved over the decades. While Collins has always been quoted as seeing the Earth as fragile, today he wants to remind people that he had two very different views of Earth, producing entirely different perspectives. One view was from his low earth orbit vantage point on Gemini 10 and the other from the moon. "You don't think of Earth as being a small or fragile object from low earth orbit," says Collins. "It's a substantial-looking hard rock. From the moon, smallness and fragility are what came to my mind because it was hanging out in the cosmic void all by its lonesome."

"It all depends on your definition of fragile," said Bean. "Earth's been around, what, four billion years or something, so it can't be too fragile." But Bean offered an important qualifier that lines up with Collins and the rest of the Eagles who believe that man's treatment of the planet is straining what's called its "carrying capacity," its ability to adequately sustain and feed the 7.4 billion people who live on it. "The bad things that we human beings do to the Earth," says Collins, "indicate that there is a great deal of fragility to it in terms of what happens to us."

"We humans are responsible for ourselves," Ed Mitchell said shortly before he died in 2016, "and if we're endangering our future, then we've got to learn how to do it differently and go forward into a sustainable period. Right now that seems very difficult to see how it's going to be done, but we've got to work on it."

"I'm an optimist," says former shuttle astronaut Tom Jones. "So, I actually think that our technology and our ability to apply it to problem solving will save us in the end. You know, we've been confronted with disasters all throughout our evolutionary history and we've not been done in yet, and I think we're very adaptable creatures and we'll find a way to apply the tools we have to save ourselves. So, I think switching over to non-carbon energy, nuclear, and then the renewables, where they can play a role, will get us out of the current box that we're in."

Since they are explorers by nature, astronauts believe it is in man's evolutionary DNA to one day colonize the universe. Their aims and timetables vary. Some cite man's hardwired necessity to explore and seek new knowledge. Others cite a more Darwinian calculation, as expressed by Al Worden, who has spent his retirement years raising funds for the Astronaut Scholarship Foundation (ASF): "What is the prime imperative of every living thing on Earth?" he asks. "Survival! Whether it's grass, trees, animals, people—we're no different. Our prime imperative is survival of the species, and that's why we have a space program. In my mind there's a genetic drive we have that says you've got to develop the capability to go somewhere else when you cannot live here anymore."

When the space shuttle program ended in 2011, space exploration stalled since NASA no longer had a way to get American astronauts off the Earth into space. Consequently, it has relied on the Russians (at a cost of $80 million per US astronaut) to get to the ISS, where the only human activity in space is still happening.

That situation may change soon as a handful of mostly American entrepreneurs have emerged to help NASA build new rockets and spacecraft that are worthy successors to Apollo's incredible machines. There is one tech mogul in particular the Eagles are betting on, and they point to his company's motto, *Gradatim Ferociter*, as the reason they believe in him. It's the same bold but calculated risk-taking strategy they say got them to the moon—*step-by-step ferociously*.

Chapter Eight

GRADATIM FEROCITER: THE FUTURE OF SPACE TRAVEL

Aerospace company Blue Origin's coat of arms
(Courtesy of Blue Origin)

- ➤ Cooperate on a planetary scale
- ➤ Don't be risk averse
- ➤ Step-by-step wins the day
- ➤ Trust your inventiveness and America's entrepreneurial genius
- ➤ Learn from the past to peacefully explore the universe

Slow is smooth and smooth is fast and we're not going to stop.
We're just going to keep going.   *Jeff Bezos, Blue Origin*

A base on the moon is a very important linchpin if we want to
be a spacefaring species.   *Harrison Schmitt, Apollo 17*

The pairing of words, how they're mated and juxtaposed, brings
to mind Mark Twain's instructive comment about word choice:
"The difference between the almost right word and the right word
is really a large matter. It's the difference between the lightning bug
and the lightning."

Lightning didn't strike immediately for the wealthiest man in
the world. He wanted to find two words that would capture the
scalability of the mission he's dreamt of since high school but that
would also demonstrate the incremental discipline required to
accomplish it. "I came up with *Gradatim* first and I wasn't happy
with it, and so I played with it for a long time and finally added
the *Ferociter*," says Jeff Bezos, to show the resolute tenacity of his
aerospace company, Blue Origin. *Gradatim Ferociter* is Latin for
step-by-step ferociously.

"Spaceflight is a high-risk business," said Bean, "and when I
first heard his motto, I thought he's got it just right." What lived in
Bean's memory, and the fraternity of all astronauts, was the ones who
died. There are the two shuttle disasters (*Challenger* and *Discovery*)
that killed fourteen astronauts in 1986 and 2003, the four men the

Soviets lost on Soyuz missions in 1967 and 1971, and the Apollo 1 launchpad fire that killed its three-man crew.

Reliability and *quality* remain the cornerstones of NASA's engineering culture. It was one of the boxes in Mueller's five-box management structure with 99.9 percent reliability the engineering objective. The lives of the astronauts depended on it. And while the Eagles were prepared to take calculated risks, they weren't grandstanders, or daredevils, but highly trained engineers and test pilots who knew their machines' limits.

It's that delicate mix of boldness that President Kennedy spoke of, combined with the Eagles' conviction, as noted by Dr. George Ruff, that risks could be minimized by thorough planning and conservatism. NASA progressed from Mercury to Gemini to Apollo, step-by-step, like Bezos's motto, ensuring that each building block created a solid foundation for the next spacecraft being developed. "Slow is smooth and smooth is fast," Bezos likes to say. "Skipping steps doesn't get you there faster. It's an illusion."

When asked to judge the current space race between the private space titans, the Eagles cite *reliability* and *quality* as their major concerns. They are impressed with Elon Musk's SpaceX, which, at the moment, is the clear leader by number of launches and contracted NASA missions, but they find themselves more philosophically aligned with Jeff Bezos, who they believe is a more reliable long-term bet because of his step-by-step approach, and his personal fortune, which gives him the staying power to keep funding Blue Origin.

"Bezos has really committed himself and his gains from Amazon to do something he thinks is very important," says Harrison Schmitt. "Musk is a very different creature, I wouldn't call him a salesman so much as a promoter, who has been supported by government funding, whereas Bezos is not."

Bill Anders worries that Musk is underestimating safety issues such as radiation mitigation in his Mars plans. Anders had been trained as a nuclear engineer, and one of the reasons he was picked for Apollo 8 was to measure the effects of radiation on the command module as it passed through the Van Allen radiation belt on its way to the moon. Since it was man's first extended voyage outside Earth's protective magnetosphere, no one knew if the radiation in the 230,000-mile cislunar corridor to the moon would be a problem. Anders's onboard Geiger counter found little penetration of the CM, but he says this would not be the case on a nine-month journey to Mars, where much stronger galactic cosmic radiation might be lethal to the crew without new technologies to make the spacecraft radiation-proof.

Excited about his Mars plans, Musk invited Anders to address the board of SpaceX. But Anders says he was promptly disinvited when he cautioned one of Musk's top aides before the meeting that SpaceX's timeline for getting there in 2022 was wildly premature because Musk had failed to provide a solution for deadly radiation exposure.

Buzz Aldrin, who advocates colonizing the Red Planet, also believes Musk has failed to think through the multiple problems and challenges of living on Mars, where radiation levels are so high it would require human colonists to live underground for significant periods of time. "You have got to live in something," says Aldrin. "You have to prepare for all of that."

In his post-NASA career, Aldrin has become a trusted and respected voice consulted by presidents, and others, regarding future space exploration. He has focused on the different strategies of the aspiring new space explorers, who beyond Bezos and Musk amount to only two other serious contenders: Virgin Atlantic's Sir Richard Branson and former Microsoft cofounder Paul Allen, who died in 2018. Allen built Stratolaunch, the largest airplane in the world—with a wingspan twice that of a Boeing 747. It's configured to launch rockets into low earth orbit from its massive fuselage. "Our Stratolaunch carrier aircraft," said Allen, "will take off from a runway and fly to the approximate cruising altitude of a commercial airliner before releasing a satellite-bearing launch vehicle. As the launch vehicle rockets into orbit, Stratolaunch will fly back to a runway landing for reloading, refueling and reuse." The targeted debut of Stratolaunch's full capabilities is 2020, but as of this writing it has no customers for payloads of any sort.

Branson's vision is far less lofty, less complicated, and with a possibly quicker payoff—should it prove safe enough. He's betting that tourists are willing to pay $250,000 to ride in his *Unity* space plane, which can carry six people fifty thousand feet above the Earth, to the edge of space, where they'll experience a few minutes of microgravity and then land back on Earth. Even though the plane's predecessor, the VSS *Enterprise*, blew up and killed its pilot in a 2014 test over the Mojave Desert, seven hundred people have reserved seats on the *Unity*, including celebrities Brad Pitt and Katy Perry. Tests of the suborbital space plane in early 2018 proved successful, and Branson hopes to have the first tourists up within a year.

With this remarkable renaissance in rocketry development by private businessmen, and the three successful launches of SpaceX's Falcon rockets in 2017, Aldrin decided that it was time to formally recognize their achievements with an annual Space Innovation Award through his ShareSpace Foundation. Aldrin was marking what amounted to a passing of the torch from the Eagles (and the old NASA) to a new generation of explorers who weren't riding the rockets but building them. They are the ones generating the news stories now with their well-funded ambitions of interplanetary exploration. They are the ones that books are being written about. How telling that it would be Aldrin to celebrate this growing shift to the privatization of the space race—once the exclusive domain of NASA and the Soviets. His choice for the first award was Jeff Bezos: "I often tell people that my middle name is Innovation," said Aldrin, who was joined at the awards ceremony by Mike Collins, "so I pay attention when people are doing innovative things. Jeff Bezos told me on a recent visit to Blue Origin that he has been dreaming of space since, at the age of five years old, he watched Neil, Mike, and I journey to the moon during the Apollo 11 mission in 1969. Since then he's charted his course through innovation, and he's been quietly breaking barriers with Blue Origin."

Bezos, who shook the hand that touched the moon, says it was one of the greatest honors of his life: "You know he's a childhood hero as are a lot of these astronauts. They and the engineers behind Apollo are what inspired me to go into engineering and science and follow that path."

Jeff Bezos, founder of Blue Origin, inspects New Shepard's West Texas launch facility before the rocket's maiden voyage *(Courtesy of Blue Origin)*

Aldrin's fellow Eagles, who gave him the nickname Dr. Rendezvous, credit him with helping pioneer orbital rendezvous mechanics (the technique of docking the command and lunar modules), without which the moon landings would have been impossible. He has a fervent mind that never stops humming with new ideas about interplanetary travel. But Aldrin's place in history, as the second man to walk on the moon, is a study in the agony of being "number two." Years ago, Bean said he listened sympathetically to Buzz talking about the impact it had on his psyche. But he couldn't resist a little levity, "and I told him that if I'd been second off the ladder, I might have been just as f***ed up as you are." Fortunately, Bean said, Buzz took it the right way and they both had a good chuckle.

Aldrin has always denied any resentment for not being chosen over Armstrong for that immortal first step. What he freely confesses to, however, in his books and lectures, are his dual demons of alcoholism and depression, which he attributes partly to his mother, who committed suicide the year before he landed on the moon. He has been divorced three times, and has pursued the limelight (the *Dancing with the Stars* TV show) in ways the Eagles find distasteful. What they continue to applaud is his ShareSpace Foundation, which provides schoolchildren around the country with educational materials to inspire deep space exploration.

MAKING SPACE TRAVEL AFFORDABLE

When Bezos received Aldrin's Space Innovation Award, Blue Origin had achieved a historic first: in 2015, it successfully landed a reusable rocket (twenty-eight days ahead of SpaceX). Blue's New Shepard rocket, named after astronaut Alan Shepard, ascended sixty-two miles to the Kármán line (the boundary between Earth's atmosphere and outer space) and descended with feathery grace back to Earth to make a pinpoint landing. Except in science fiction movies, no one had witnessed a booster rocket deploy landing gear with four sprouting legs, sliding down like a giant spider on a silky thread, to a full stop on a dedicated landing pad.

Blue Origin's New Shepard space vehicle successfully flew to space before executing a historic landing back at the launch site in West Texas on November 23, 2015 *(Courtesy of Blue Origin)*

In the annals of space rocketry, this was a technological paradigm shift with long-term economic implications. In the budget-rich Apollo era, the Saturn V rocket was designed to be disposable. Its massive first-stage booster, powered by its incomparable five F-1 engines (each measuring twelve by eighteen feet) would jettison from the command and service modules and drop off into the ocean, never to be seen or used again. On a smaller scale it was the equivalent of driving your car cross-country, *one time*, and then dumping your vehicle (engine and all) on the side of the road only to build a new one each time you wanted to drive again.

Today, at the bottom of the Atlantic Ocean, sit thirteen Saturn Vs, each of which cost NASA $1.16 billion to build in today's dollars. That represents a total of $15 billion dollars in discarded hardware now rusting on the seabed with only a onetime return on investment.

While this made little business sense, Bezos understood that in the 1960s von Braun's engineering team had neither the time nor the technology to create a reusable rocket. But today's supercomputers, he says, can create rocket designs faster and better and with the kind of precision never before possible. By 2000, Bezos was sensing that what the Internet and computers had done for the business world—and made Amazon possible—the same tectonic shift could shake up the space industry, launching it into a new golden age. He was ready to bet big again. Even before he founded Amazon in 1994, Bezos says his first passion, his real goal in life when he graduated from Princeton in 1986, was to build an aerospace company that would create the architecture and infrastructure for systematic planetary exploration. He had been envisioning it since high school, when he ended his valedictorian speech by saying, "Space, the final frontier, meet me there!" It wasn't just a cute throwaway line or idle *Star Trek* talk for Bezos. It was the first public hint of an almost messianic calling sparked by the Apollo moon landings, to turn humans into an interplanetary species.

The historical scope of his passion for all things Apollo surfaced in 2013 when he funded an expedition to find Apollo 11's five F-1 rocket engines, which had catapulted Armstrong, Aldrin, and Collins on their 240,000-mile trajectory to the Sea of Tranquility. Like a digital age treasure hunter, Bezos says it took him all of fifteen minutes on the Internet to discover NASA's radar-tracked coordinates of where the booster stage of Apollo 11 had impacted the surface of the Atlantic Ocean. He was transfixed with getting his hands on those legendary engines. He hired a sixty-person team, and after three weeks of using state-of-the-art deep-sea sonar, his recovery crew found the rusted F-1s sitting nearly three miles below the surface (1,500 feet deeper than the *Titanic*'s resting place). Serial numbers proved they belonged to Apollo 11. Bezos had just time traveled back fifty years to the moment of Armstrong and Aldrin's liftoff, staring at the engines that hurled them into the arms of destiny. It wasn't just curiosity, or ego, driving him. He felt a powerful obligation to history to preserve the F-1s for current and future generations. They now sit in Seattle's Museum of Flight for public viewing.

What is significant, says Bezos, is that "the F-1 engine is still a modern wonder," so much so that neither NASA nor private contractors nor the Russians nor anyone else has a rocket today to match the Saturn's heavy lift capability. That its superior engines should be on display in a museum, instead of flying again in some updated version, says much about the state of space exploration today and what brought us to this point.

The harshest assessment comes from Anders, who says, "Our space program has been on the wrong track ever since the end of Apollo with the Space Shuttle. The Space Shuttle, as spectacular as it is, it's a dead-end program and that's certainly brought to point by it being terminated, and now we have to hitchhike rides with our prior competitors, who must be laughing up their sleeves that the Americans, who got to the Moon first, now have to ride on Russian rockets."

Anders knows the shuttle's backstory intimately because he reluctantly helped green-light it when he served from 1969 to 1973 as executive secretary for the National Aeronautics and Space Council. It was his job to help the Nixon administration figure out what NASA's post-Apollo future would look like. He was witness to the fateful political decision fifty years ago to shift NASA's vision and operational profile to Earth-centered missions focusing on communication satellites, weather, and scientific and human experiments in zero gravity. Astronauts would no longer be reaching for Mars or leaving Earth again for any other planet in the solar system. The carte blanche funding was over, and Anders and all the Eagles saw it coming as soon as Kennedy's moon shot was accomplished. Apollo-level budgets, they knew, were unsustainable.

Once the race was won, not only did presidential and

congressional largesse start fizzling, but so did the public's interest in space. By Apollo 13, the TV networks had even stopped giving live-broadcast coverage of the crew's onboard television show, as it had with Apollo 11 and 12. Lunar fatigue had set in. Television screens were now dominated by the Vietnam War, which had thrown a hand grenade in the middle of America's political and cultural landscape. War coverage, race riots, and earthly crises were replacing moon coverage on the nightly news broadcasts and the rest of the media. Apollo had inadvertently returned man's focus to Earth. The moon had become old news and no longer had anything new to offer a nation turning inward with political and social upheaval.

When it came time, therefore, to rethink NASA's mission, former NASA chief historian Roger Launius says the agency reached back to the blueprint of its most brilliant architect: von Braun's original 1950s *paradigm* for space exploration, which included his idea for "a large-winged, reusable spacecraft to make space access routine." And thus the space shuttle was born. President Nixon supported the shuttle, in part, because NASA believed it could save money. Anders says that, of the two shuttle designs proposed, he advocated for the smaller version because it would be less expensive to build and easier to test NASA's assumption of a tenfold reduction in the cost of space missions. Since most of the shuttle would be built in California, it came down to one thing: "Jobs," says Anders. Politics was about to overrule financial prudence and NASA's long-range health. "I got a call from Nixon's chief of staff,

H. R. Haldeman, and all he wanted to know was which version of the shuttle would generate more aerospace jobs. It was obvious the larger shuttle would and that's what they picked."

During their thirty years in operation, the five reusable shuttles (also known as orbiters) flew 135 missions. They facilitated the deployment of the Hubble Space Telescope (whose images revolutionized our understanding of the universe), helped in the construction of the ISS, launched numerous satellites and interplanetary probes like Voyager 1, and conducted extensive science experiments in orbit. "It did a lot of wonderful things but at extremely high cost," says Anders. "The space shuttle was supposed to cost one-tenth the cost per pound to orbit than the Saturn V modified. Instead, it cost ten times that in retrospect. That's a 100-fold error, so the space shuttle had eaten NASA hollow from the inside out."

None of this was lost on Bezos, who could see that the shuttle's late-1970s technology had, over its life span, saddled it with prohibitively high maintenance costs and aging hardware, which undermined safety and negated the intended cost benefits of its reusable orbiter and semi-reusable booster. "I think with hindsight as a guide, they got that decision backwards," says Bezos. "They should have made the orbiter expendable and made the booster a fly-back reusable booster. We would be living in a very different world now, I think, if they had taken that path."

While observing NASA's struggles, Bezos quietly began studying the business model of space with the same hard-nosed analysis he applied to e-commerce and its scalability. He suspected reusable rockets could radically alter the cost of getting to space (possibly by a factor of ten) and make it profitable in the long run. With the enormous financial cushion of his "Amazon lottery winnings," as he likes to refer to his multibillion-dollar success, Bezos finally pulled the trigger on his true ambition. He decided to start small and stay under the radar. In 2000, without any public announcement, he became the first of the space billionaires to launch an aerospace company (two years ahead of SpaceX) with just six employees, mostly physicists, in a space not far from Amazon's headquarters in Seattle.

Bezos says he spent the first three years working in secrecy with his small team of rocket scientists exploring novel launch technologies that might surpass existing chemical rockets. Always looking for disruptive technologies, Bezos wanted to make sure that the rocket pioneers, like von Braun, hadn't missed anything. They hadn't. After exhausting all the alternative possibilities, Bezos came to the conclusion that chemical rockets actually were the best solution for an Earth launch.

"And so in 2003," says Bezos, "I reconfigured the company very significantly to focus in on reusability." He began hiring engineers from the shuttle program and members of the McDonnell Douglas aerospace team, which in the 1990s was testing a scaled-down reusable rocket for NASA and the Department of Defense. Known as the Delta Clipper Experimental (DC-X), the diminutive thirty-nine-foot rocket flew a mere 150 feet above the ground in its first flight and successfully relanded. While it looked promising, the funding for a larger orbital version of the DC-X dried up, and NASA canceled the program. But Bezos and his engineers recognized the missed opportunity and pursued reusability step-by-step, quietly and slowly, like the tortoise, which is Blue Origin's mascot and is featured in—what is unusual for any modern company to have—a coat of arms. Coats of arms date to the twelfth and thirteenth centuries, when royal families used their heraldic visual designs of crests, with mottos, as symbols of nobility and strength. There is nothing space age about them, and no other aerospace company has one. Bezos, however, saw a coat of arms as a competitive differentiator to motivate his engineers and tie Blue Origin's space mission to its earthly origins (the company's name refers to the blue planet, Earth, as the point of origin). The design is loaded with instructive symbolism and clues to who Bezos is as a businessman and a student of history. It evokes the era of the great fifteenth-century explorers, showing sailing ships traversing the oceans, while in the background two tortoises point toward the heavens as a rocket blasts off into the cosmos. At the base of the Earth sits a winged hourglass embedded in a scroll with the words *Gradatim Ferociter* in bold letters.

Since time and money are inexorably linked, Bezos put the hourglass at the bottom of the coat of arms, which he says is a Victorian cemetery symbol signifying that "time is fleeting." "We don't have forever," he cautions.

The tortoise mascot is a reminder of Aesop's fable that the tortoise, through ingenuity and perseverance, eventually overtakes the faster, more arrogant hare. "If you're building a flying vehicle, you can't cut any corners," says Bezos. "If you do, it's going to be just an illusion that it's going to make it faster. You have to do it step by step, but you do want to do it ferociously."

And the company's feather logo is simple, Bezos says. "It's just a symbol of the perfection of flight. For thousands of years, we humans have been looking up at the birds, and wondering what it would be like to fly....I think it's representative of freedom and exploration and mobility and progress."

Bezos is also fond of repeating the Navy SEALs slogan "Slow is smooth and smooth is fast"—something that Navy SEAL and astronaut Chris Cassidy, who flew the shuttle, says is applicable in space exploration. "My personality is much more in line with how Bezos runs his business. Just getting it done, head down, with a really sound team. Not to take anything away from Musk, but they're much more about the show. He keeps the public excited with successful launches, but they're not ready to put *people* on a rocket which is what they're ultimately contracted to do and what NASA is providing them money to do. NASA's not providing them money to put a big PR campaign with a guy in a Tesla and a spacesuit. So, I really like the way Bezos is going about it a lot."

Nicole Stott echoes Cassidy's sentiment that launching a red Tesla into space may be good theater but runs counter to the inherent aversion astronauts have to arrogance and showiness: "I love the way Blue Origin is doing it," says Stott. "I mean they release information about what they're doing as they think it's worthwhile, or it's another milestone in what they're doing. It's not about the show."

SpaceX views the "showmanship" criticism as unfair and missing the larger point of its lengthy track record of successfully launching fifty missions for NASA and other government contractors. SpaceX's senior communications manager, James Gleeson, says launching Musk's red Tesla into space with its "Starman" mannequin strapped to the driver's seat actually had a Kennedyesque effect of reigniting the public's fervor for space. "We used the opportunity to put a payload into space that would inspire people, as opposed to just putting a dummy weight or a block of concrete up there. Who wants to see that?"

Starman, which is still on a trajectory headed for Mars, generated 2.3 million YouTube views and worldwide news coverage. "A lot of people [wondered] what's the purpose of sending a car to Mars? There's no point, obviously!" Musk told CBS News after the launch on February 6, 2018. "It's just for fun and to get the public excited." Critics noted it also happened to be great advertising for Tesla, Inc., which at the time was struggling with production problems and on the verge of laying off four thousand workers.

Like Bezos, Musk is a transformative figure who has deconstructed traditional business models and exceeded Wall Street's expectations. He was born in South Africa in 1971 and taught himself computer programming when he was twelve years old. He survived childhood bullying that was so severe he wound up being hospitalized after being thrown down a flight of stairs. After graduating from the University of Pennsylvania, he went to Stanford University to pursue a PhD in applied physics but, restless to pursue various entrepreneurial ideas, dropped out after two days. His biggest triumph came as the cofounder of PayPal, which was bought by eBay in 2002 for $1.2 billion. Although he was too young to remember the Apollo missions, Musk was drawn to Mars, where he wanted to air-drop a miniature experimental greenhouse containing food and crops to see how it would adapt to the Martian environment. When he couldn't get anyone, including the Russians, to sell him a reasonably priced rocket to get there, he dove into the costs of rocket manufacturing and decided he could build one cheaper himself. What was also driving Musk was a dystopian anxiety that humanity's time on Earth was running out, that with one roll of the cosmic dice, the planet could be wiped out by an asteroid or by man's own destructive hand through nuclear war, climate change, or some biological catastrophe.

It is a fear shared by Anders and some of the other Eagles, who agreed with physicist Stephen Hawking's final prediction before his death in 2018 that man has less than a thousand years left on Earth. "I'm not sure we have even that much time," says Anders, who at this stage in his life describes himself as a "pessimistic idealist."

"With climate change, overdue asteroid strikes, epidemics and population growth, our own planet is increasingly precarious," Hawking said. He urged man to begin colonizing another planet within the next hundred years or face extinction. Al Worden agrees: "Too many people want to wipe each other out; there's too much hatred" for man not to look for an escape plan to other planets.

Musk had long been convinced that space migration was the home planet's ultimate defense and destiny, and thus leveraged his PayPal fortune in May 2002 to launch SpaceX. His analysis of the space industry led him to the same conclusion as Bezos: that rocket reusability could drive down the cost of production by a factor of ten. The timing for private space initiatives was good. In 2004, President George W. Bush announced the planned retirement of the space shuttle and supported a new plan called Constellation to return man to the moon with the ultimate goal of a manned flight to Mars.

The problem was money. Bush was giving NASA a limited budget to pull it off, and that set in motion the agency's first outreach to private entrepreneurs for low-cost rocket launch alternatives. In short order, they awarded a $227 million, no-bid contract, in 2004, to Kistler Aerospace, whose CEO just happened to be revered NASA veteran George Mueller. When he heard the news, Musk was outraged by the contract because Kistler owed creditors $600 million and was struggling to survive. Sensing in-house favoritism, Musk did something everyone told him not to do if he ever wanted future business from NASA: he sued the agency. And that wasn't all. He took his fight public by going to Congress with the backing of Citizens Against Government Waste to complain that NASA was cheating taxpayers by awarding the Kistler contract without a single competitive bid. The Government Accounting Office, overseeing the process, agreed with SpaceX, and NASA was forced to cancel the contract. SpaceX was on its way to leveling the playing field for scrappy, upstart rocketeers who knew they could build rockets at lower cost than the old boy network of aerospace contractors who had ruled since the Apollo years.

Over the next fourteen years, SpaceX slowly convinced NASA and private contractors that its Falcon rockets (named after the *Millennium Falcon* spaceship in *Star Wars*) were reliable enough to launch satellites and take supplies to the ISS. Even though one of its Falcon 9 rockets blew up in 2016 carrying an Israeli satellite, which would also be used by Facebook, Musk's Falcon Heavy rocket was certified by the Air Force in 2018 to carry a classified AFSPC-52 satellite in 2020.

What the Eagles, NASA, and the current class of astronauts want right now is to stop relying on the Russians to get to the ISS. SpaceX and Blue Origin have proven that their new generation of reusable rockets are viable for sending nonhuman payloads into space at an affordable cost. What has yet to be demonstrated is their ability to carry humans into orbit and beyond as safely as the Apollo command modules did in the 1960s and '70s. This is what the space community is waiting to see, and it is why Bezos says he's been moving like the tortoise to make sure he can deliver a human-rated space capsule that the astronauts can believe in. It is the main reason they have so much faith in his step-by-step approach.

Going forward, the larger question hanging over NASA today is: What is its goal? Not one president since Kennedy has articulated an American vision for space with the same grandeur, purpose, and commitment as the moon shot. Where, exactly, do we want to go? Where's the money? Where is the continuity from one administration to the next? asks Tom Stafford, who descended within nine miles of the lunar surface on Apollo 10 to test the lunar module in preparation for Aldrin and Armstrong's landing. "I'd like to see continuity in this administration because we've had a lack of continuity over several administrations." Each recent president keeps changing the mission of his predecessor, Stafford complains. President Obama canceled George W. Bush's Constellation program to return to the moon as "been there done that." Obama envisioned crewed missions to an asteroid, Mars eventually, and accelerated the privatization of space. President Trump, in turn, killed Obama's plan and decided again to go back to the moon per President Bush, but asking for an even bigger assist from the private sector.

"This time, we will not only plant our flag and leave our footprint, we will establish a foundation for an eventual mission to Mars," Trump said as Schmitt watched him sign Space Policy Directive 1 at the White House on December 11, 2017.

"I think it's a first step," says Schmitt, "but they didn't recommend an implementation strategy, so we'll have to wait and see what the Space Council and NASA's new administrator will do."

"It's all hot air until someone actually does something," says Stephen Kane, an exoplanet scientist, about Trump's plan. Kane, a researcher at the University of California, Riverside, is a lead-ing expert in planetary habitability, and like many scientists in the space community, he's eager for more deep space probes looking for life and manned voyages to pick up where Apollo left off in 1972.

Most frustrated of all are the Eagles and the forty-four astro-nauts who are currently on NASA's active roster and have nowhere to go but the ISS. Soon, even that destination might vanish as NASA's new administrator, Jim Bridenstine, announced that US funding for the ISS will cease in 2025. He wants most of its opera-tions turned over to private spaceflight companies. But no compa-nies have shown interest in taking it over, and NASA's ISS partners, Russia, Canada, Japan, and the twenty-two nations that make up the European Space Agency (ESA), have yet to offer a plan for its continued use. China, sensing NASA's irresoluteness, is moving to fill the void by offering all members of the United Nations the chance to work on its own space station, scheduled to begin opera-tions in 2022. Schmitt worries that China will displace America as the leader in space exploration: "And if this administration and the Congress and the media don't recognize the geopolitical threat represented by the Chinese, then it will be China that dominates deep space."

Bridenstine, who passed Senate confirmation by only two votes because of his early doubts about climate science (he now believes in climate change), has yet to lay out specific goals with deadlines, according to NASA insiders, let alone an implementation strat-egy. His one consistent message has been Trump's: that private enterprise must play a key role in its plans for the space program and return to the moon.

This lack of deadlines, focus, and funding has not deterred the Eagles from advocating what has become the most widely supported strategy for learning how to live on other planets. It is an incremental one. It has buy-in from career NASA staff and its astronauts, including the space agencies representing Europe, Canada, Russia, and China. They all see the moon as the logical gateway to the universe, a training ground for astronauts, and an eventual staging area for Mars and then the rest of the solar system. Here's why: the moon, we have now discovered, has all the resources (water, oxygen, silicon, titanium, iron) to create a moon base that can—and this is the critical part—easily be resupplied from Earth because it's only a three-day journey.

"Many of us see a return to the moon to develop a permanent moon base," says Charlie Duke, "to learn how to extract the minerals from the moon and to set up moon-based telescopes and other kinds of instruments, and then develop the systems where we cycle crews back and forth like we do on the space station, every three months to six months."

"I think we need to go back to the moon building up the infrastructure and the architecture, and you have to do it step-by-step before thinking about Mars," says Jim Lovell. "The moon is a very, very helpful place if you want to go anywhere else in the solar system," says Harrison Schmitt. "You can minimize the mass required to get to Mars, by using lunar resources, not only water but oxygen and hydrogen and potentially even food supplies. So the moon is a very important linchpin if we want to be a space-faring species."

"What we've learned from working on the Space Station," says Nicole Stott, "is that you can build really complex things in space, and if we get back to the moon, we can do complex things there that then make it a heck of a lot easier to get to Mars than launching some big rocket off the Earth to get there."

A UTOPIAN VISION

When Bezos looks at the moon, he sees a giant laboratory for advancing man's knowledge and a staging area to build spacecraft for further exploration of our solar system. He actively disputes Hawking's and Musk's dystopian views that humanity's Earth days are numbered and that space migration is the Plan B to escape its inevitable catastrophes.

"Instead, there's this other vision of the future," says Bezos, "which I do want my grandchildren to live in, and their grandchildren, which is a trillion humans in the solar system. Because if you have a trillion humans, you'll have a thousand Einsteins and a thousand Mozarts, and we'll have this incredible civilization. I want Plan B to make sure Plan A works, and I assure you," Bezos says (like the Eagles), "that Earth is the best planet in the solar system, and that we go to space to save Earth."

It can all be done, says Bezos, and that's why he's been liquidating $1 billion of his Amazon stock every year to fund Blue Origin. He's already in talks with NASA about manufacturing lunar landers that can one day supply lunar colonies. Solar energy, he says, can be a primary source of power on the moon, as envisioned in a 2017 proposal by the ESA. The ESA wants to create lunar outposts near what are called the eternal peaks of light along the moon's polar mountains, which, because of their location, are in nearly constant sunlight. What this means is an endless source of solar power to fuel a lunar colony, which can be expanded over a multi-decade effort and lead to permanent human settlements. Lunar greenhouses would be set up to grow food.

Back on our home planet, Bezos says our unrestrained energy consumption will eventually require space-based solar power and other energy sources. "If you take the baseline energy usage on Earth today, and just compound it at a few percent a year for just a few hundred years, you would have to cover the entire surface of the Earth in solar cells. And that's obviously not going to happen." In order to meet those needs, Bezos foresees a day when Earth is zoned residential...meaning heavy manufacturing would be moved off planet, giving Earth's biosphere a chance to replenish itself. Solar power plants, for example, could be built and put in orbit near the sun, where they could harvest 24/7 sunlight and beam it back to Earth.

On paper, and in reality, scientists say moon bases, mining asteroids, and orbiting solar power plants are possible. But when?

Most of it will not occur in his lifetime, says Bezos. What is possible, he believes, is to have a Blue Origin lunar lander completed and on the surface of the moon by 2023. Musk predicts his BFR rocket, which stands for Big F***ing Rocket, will launch a cargo mission to Mars in 2022 in preparation for a Martian colony.

The Eagles are cheering these bold efforts but want to remind the world of what they realized half a century ago when they visited the moon: humans must see themselves as one race, which will explore space as Earthlings instead of separate nationalities. How, the Eagles wonder, will nations behave on the moon when precious minerals and gases like helium-3 (He-3) are at stake? Might a new destructive gold rush be unleashed by He-3, which is rare on Earth but abundant on the moon? He-3 is valued at a head-spinning $3 billion per ton (gold is valued at only $64 million per ton). The reason for He-3's high valuation is that in large quantities it could finally make the holy grail of fusion power achievable and replace today's far less efficient nuclear fission reactors. The Chinese reportedly have plans to set up mining operations on the moon to extract He-3, and the Russian firm Energia has shown interest, too.

While the Eagles planted six flags on the moon during their missions, America abided by the outer space treaty of 1967 signed by 106 nations, which prevents anyone from laying claim to any part of outer space, or its celestial bodies and the moon. It was followed up in 1979 by the moon treaty, which said that any use of the moon should be for the benefit of all mankind. It is not encouraging that the US, Russia, and China did not sign that treaty. Then, in 2015, both houses of Congress passed the SPACE Act, giving American citizens the right to extract resources from outer space and other worlds and sell them for personal gain. What this means is that the moon is open for business. We can only hope that nations will treat it as a common ground for cooperation and scientific advancement.

When the European explorers returned from the New World in the sixteenth and seventeenth centuries, they did not leave plaques behind, as Armstrong and Aldrin did on the moon, which read, "We came in peace for all mankind." Instead, the major superpowers of the day, Spain, Great Britain, France, and Holland, gradually returned to the Americas to carve it up in murderous land grabs. Does the same fate await the mineral-rich moon and the rest of the universe? Neil Armstrong, who was rarely given to hyperbole or pessimism, pondered this issue and offered a disquieting assessment of man's future that should be read and reread. At the Starmus Festival in Tenerife in 2011, the year before he died, he told his fellow astronauts and scientists (with a slight tremor in his voice) that he did not believe man had evolved enough as a species to colonize space: "Based on our record here on Earth, we are not yet qualified to populate and govern a larger segment of our universe. We may or may not have time enough to grow as a species to control our ultimate destiny."

Armstrong, however, did not want to end on an unhopeful note. With the same faith he summoned to steer the Eagle clear of lunar boulders to a safe landing, he offered the audience faith that man might be able to change:

Yet there is great reason for hope. And we have no other choice. There is no doubt that our instincts will force us to try. On reflecting on the same subject two millennia ago, Plato, around 400 BC, said: "We must take the best and irrefutable human doctrines and embark on that as if it were a raft on which to risk the voyage of life," which indeed it is and indeed we must.

Chapter Nine

THE NOBLEST JOURNEY OF ALL

A mother eagle nesting near the Kennedy Space
Center in Florida

KEY LESSONS

- Don't spin the truth, and call out bullshit
- Beware of false patriotism and inflated heroism
- Maintain your honor and integrity

- Put moral courage, country, and the public good first
- Pursue your passion and you will thrive
- Be understated and humble in all your endeavors

"Mister President, this is wrong," I told him! "You weren't responsible for it. This was a Democratic program, and you just came into office."   *Frank Borman*

Moral courage is standing for something you believe in in the face of this big opposition, so that takes a lot of sustained courage to stand your ground.   *Charlie Duke*

Moral courage is a rarer commodity than bravery in battle or great intelligence.   *Robert Kennedy*

Frank Borman says he would like to be remembered "as a committed Christian who dearly loved his wife, his family, and his country, who always strove to do his *duty* and never compromised his honor or integrity to achieve a goal." Part of not compromising his honor and integrity, Borman says, included standing up to President Richard Nixon over the White House's preparations for the moon landing. Borman's backstory at the White House highlights the less visible but nobler journey of moral courage the Eagles traveled—a virtue they hope Americans will always strive for.

Nixon took office on January 20, 1969, six months before Armstrong walked on the moon. NASA had assigned Borman as its liaison to the White House as it came into the final stretch of that remarkable day. It required a lot of political preparation and public relations work. What would Nixon say when they landed? Would he politicize it? Would he make it a purely American moment, or one for humanity? That it all went so smoothly in the end belies how easily it could have gone the other way.

The drama began when NASA sent suggested talking points to the White House, most of which Borman says he rejected because they were politically motivated to give Nixon credit for the landing, even though Kennedy was the acknowledged father of the space program. Borman, who was not intimidated by anyone, refused to accept self-serving distortions of the facts or spinning the truth. It was another of the defining characteristics of the Eagles: their "low tolerance for bullshit," as Dr. Santy explained earlier.

"When I wrote Nixon a memo with my objections, he called me into the Oval Office to explain why. I put zero spin on it," recalls Borman. "'Mister President, this is wrong,' I told him! 'You weren't responsible for it. This was a Democratic program, and you just came into office. You didn't have anything to do with Apollo 11. You should assume that it's the American people who deserve the credit.'"

The original publicity plan for the landing was to have Nixon make a telephone call from the Oval Office, which included a long speech congratulating Neil and Buzz, followed by the playing of the national anthem. "I remember telling him during our meeting to keep his remarks short, and nonpartisan, to be inclusive of all nations...and then to thank the astronauts and just get off the air." As for the national anthem, which Nixon felt was important to play, "I told him it really didn't make much sense to stand the astronauts at attention for two-plus minutes and lose time from the valuable experiments they had to set up." Besides, it would smack of nationalism at a moment meant for all mankind.

Borman had no idea how the president would react, but to his credit, he says Nixon listened without protest and followed all of his advice. He limited his lunar phone call to sixty-nine seconds (252 words), kept it apolitical, and noted its significance for everyone all over the world:

Because of what you have done the heavens have become a part of man's world, and as you talk to us from the Sea of Tranquility, it inspires us to redouble our efforts to bring peace and tranquility to earth. For one priceless moment in the whole history of man all the people on this earth are truly one—one in their pride in what you have done and one in our prayers that you will return safely to earth.

In the end, Borman was impressed with the way Nixon handled Apollo 11, and respected him until Watergate. Nixon had trampled on the three hallowed words the Eagles repeatedly measured themselves by: *duty, honor, country.* For Borman it was unforgivable. "It was a real blow to me personally, because in my interaction with him he seemed good for the country. And then when I found out about all the lies and the rest, to me that was inexcusable. I had hoped he was trustworthy, but he wasn't."

Bill Anders would face his test of moral courage when President Gerald R. Ford appointed him as the first chairman of the newly created Nuclear Regulatory Commission (NRC) in 1975. Within his first nine days on the job, Anders learned that the defunct Atomic Energy Commission (AEC), which had preceded the NRC, was aware of small cracks in key safety piping at the nuclear power reactor in Morris, Illinois, and other reactors, mostly in the Northeast. "It was leaking," recalls Anders, "allowing secondary water with slight radiation in it to dribble into the reactor vessel."

At the time, the AEC (founded in 1946 by President Harry S. Truman) had been accused of lax regulation of the nuclear power industry because of what critics said was its higher priority of promoting nuclear power over other sources of energy. Anders vowed to keep the industry in line and to turn the NRC into an independent regulatory agency serious about oversight. By 1975 the US had fifty nuclear reactors online providing the country with less than 10 percent of its electricity. The AEC was hoping for two hundred nuclear reactors by the mid-1980s. (Today there are ninety-nine nuclear reactors in the US.)

When the Associated Press discovered the story of the leaking pipes in Morris on March 6, 1975, Anders told its science reporter, "I'm impressed with the margin of safety built into nuclear power plants, but our job is to insure public health and safety and we're going to take whatever action is necessary to do that." And act he did. The Morris reactor was shut down, but Anders didn't stop there. He feared there might be a design flaw in the cooling pipes manufactured by General Electric, so he gave the unprecedented order of shutting down twenty-three nuclear reactors within twenty days if they couldn't prove the pipes were safe. When the AP reporter reminded Anders that President Ford could fire him in a confrontation over reactor safety versus the country's energy needs, Anders replied, "It doesn't bother me one bit."

It was classic Anders. Like Borman and the rest of the Eagles, Anders took seriously his duty to country first. When ten of the nuclear power plants in the Northeast failed to meet his twenty-day inspection deadline, he shut them down. To make up for the power shortfall, coal-fired plants were forced to draw extra power from Canada for two days in order to meet the Eastern Seaboard's power needs. Anders's order to shut down the reactors cost the utility companies millions of dollars. "The nuclear power industry was yelling," Anders remembers now, "but I didn't back down and got no interference from President Ford."

MORAL COURAGE

For the Eagles, physical and moral courage are inseparable, but of the two they believe moral courage is the greater test and the nobler journey. "Physical courage, you respond to the circumstances you're facing in the moment," says Charlie Duke, "and you have to do that quickly, whereas moral courage is standing for something you believe in in the face of this big opposition, so that takes a lot of sustained courage to stand your ground. For instance, my wife and I, we're strong Christians and we want to take a moral stand on what the Bible says and in some respects that's unpopular, especially in the Millennial generation."

The ancient Greeks and Romans, says Harvard Classics chairman Mark Schiefsky, held moral courage in higher esteem than physical bravery even though it evoked less awe in the popular imagination. "For Plato and the intellectualists," says Professor Schiefsky, "the strength of the soul is more important than the strength of the body and we could formulate that the moral—that the psyche—is of greater import than the strength of the body."

It's a distinction that President Kennedy appreciated when he set America's sail for the moon. The physical bravery of the Eagles was a given. It was in plain sight with each of their fiery liftoffs. But the noblest journey of all, Kennedy believed, was not to the moon; it was the inner journey, the one leading to moral courage and devotion to something greater than oneself. These are the qualities the John F. Kennedy Presidential Library and Museum honor each year with their Profile in Courage Awards. The recipients are public servants "who govern for the greater good, even when it is not in their own interest to do so.... who choose the public interest over partisanship—who do what is right, rather than what is expedient."

"My father's heroes," says Caroline Kennedy, "were men and women who were willing to risk their careers to do what was right for their country."

As Robert Kennedy would add, "Few men are willing to brave the disapproval of their fellows, the censure of their colleagues, the wrath of their society. Moral courage is a rarer commodity than bravery in battle or great intelligence."

Throughout their lives the Eagles have practiced their courage and patriotism with quiet humility. What troubles them now is what they believe is a growing trend toward false patriotism and braggadocio in many of today's elected officials: "Most of the congressmen and senators serving now," Bean said, "they need to rip that American flag off their lapel, and put a picture of a donkey, or an elephant, or their own damn picture on their lapel because they're not patriots to this country. They are self-serving jackasses."

In that same vein Collins abhors the adulation of celebrities and sees it as one of America's greatest cultural failings—along with what he calls "the inflation of heroism." "Celebrities and celebrity are nonsense," he says, "an empty thing for someone to aspire to. Can you imagine a person who is known for well-knownness," quoting his friend historian Daniel Boorstin, who has published nearly two dozen books about American history and government. Collins prefers to honor the everyday heroism practiced by firemen, nurses, first responders, and others who save lives. Astronauts, he says, are not heroes. "We astronauts were good; we worked hard; we did our jobs to near perfection, but it is what we signed on to do." The men the Eagles considered heroes were their colleagues in Vietnam who were dying for a cause that, in the end, few had come to believe in: "I've always had a guilt complex to some degree," said Gene Cernan. "My buddies were getting shot at, shot down, and, in some cases captured, and I was getting my picture on the front page of the paper."

"A lot of our friends were flying planes in combat in Vietnam and there would we have been had we not been in the space program," Collins says. That is why his measure of heroism is the nation's highest award for it: "I go by the Congressional Medal of Honor, 'above and beyond the call of duty.' That phrase, 'above and beyond the call of duty,' is what matters," he says.

Today, Collins likes to describe himself as a "good retiree," who, at eighty-eight years old, stays in shape by doing mini-triathlons once a year. He finds peace and comfort in the following: his two daughters, Ann and Kate, and writing poetry. "I love poetry," he says. "I read it and quote it, but the poem I just wrote…is terrible." There is an unaffected modesty to Collins, which is a common trait of the men who went to the moon. He has an inviting handshake and warm eyes that draw you in with his curiosity about who you are, instead of who he is. His sentience is evident from his poetry. When he first saw the far side of the moon, he was inspired to write this ode, which appeared in his book *Carrying the Fire*, in 1974:

Cold stones jumbled in a heap.
Lifeless plains, sharing only the sun
With a verdant recollection I must keep,
Till I next see one:
One penny, one peony, one misty waterfall.
For me a choice—to hear a voice,
Or slip by on it all.

As a member of the first moon-landing flight, Collins could have cashed in on his unwanted fame but, like Armstrong, he pursued a life of the mind instead of the dollar. "I'm not against making money, but all my life that's never been my objective. Every time I've changed jobs, what I've wanted to do is find something interesting, so I've looked for interesting jobs rather than money-producing jobs."

Mike Collins with his daughters, Ann (left) and Kate (right), 2014, in front of the Apollo 11 command module at the National Air and Space Museum
(Courtesy of National Air and Space Museum)

THE COURAGE TO PURSUE YOUR PASSION

"Make sure you don't get moved into a place where you're doing it for the money," agreed Bean, who pursued his passion for painting. Joseph Campbell, who wrote that "a hero is someone who has given his or her life to something bigger than oneself," also believed that part of courage involves following one's bliss:

> If you do follow your bliss, you put yourself on a kind of track that has been there all the while, waiting for you, and the life that you ought to be living is the one you are living. When you can see that, you begin to meet people who are in your field of bliss, and they open doors to you. I say, follow your bliss and don't be afraid, and doors will open where you didn't know they were going to be.

It is the same message Bean delivered to college audiences except it was directly born of his experience from walking on the moon:

> The only limits for us are the limits we place on ourselves. We [humans] may be small on the cosmic scale, but we've been given a great gift in our lifetime. We can become and do things we're willing to try, and work, and plan to do! There's nothing else in the universe, except humans who can do that. We're it! Most people are wasting that great gift. Don't waste it! You're unique.

In the same way that no two people on the planet share the same set of fingerprints, Bean marveled that we each have our own special attributes, and individual potential:

There's nobody on this earth like you. Your son may look like you, your daughter, or mother may look like you, but they're not like you. There's never been anyone like you, and there never will be another person like you. Think about this when you decide what you're going to do. You've only got about fifty, eighty, or maybe ninety years...who knows? But you've got to do something with this great gift.

"God has really given us a stage," says Lovell. "Here is the stage, here is the performance, and how the play turns out is really up to us, and if you want to do something, you have to have the courage to get started."

"The crews of the Apollo missions will forever be our heroes," says Nicole Stott. "They are not just heroes of spaceflight, they are heroes for humanity. Beyond the marvel of technology and seeming impossibility, their missions to the moon were acknowledged by people all around the world as an accomplishment for all. Seeing Earth from space with their own eyes and the images they shared with all of us were a gift of understanding—the best understanding of our own reality—of the 'who and where' we are in this universe. So how do we ever really say thank you? I believe we do that by taking their lead, by never confining ourselves to the bounds of our planet, but by continuing to explore and establish our presence off our planet with the ultimate goal of improving life for everyone here on it."

The Eagles say their voyages to the moon gave them two gifts:

gratitude for the solar system's *gift of planet Earth*, and gratitude for *the gift of life* that Earth has given us. These are the two themes they have repeated consistently in their conversations for this book. When they look at the moon today, they say earthly time and earthly crises no longer loom as large as they did before their voyages.

"I think if you do something that's drastically different, like flying to the moon and coming back again," Collins says, "everyone tells you how important it is, how wonderful it is and then, by comparison, a lot of things that used to seem important, don't seem quite as much so, and I'm not saying that I'm able to face life with greater equanimity because I've flown to the moon, but I *try* to and maybe some of our terrestrial squabbles don't seem as important after having flown to the moon than they did."

"The biggest problem the world has," says Anders, "is that we can't get together. I mean even friendly people can't get together, just look at the European Union as one example. If we're ever going to explore the universe, we need to do it as a planet."

Moving forward as a planet is the hope of all astronauts. They remain a rare and optimistic breed in a fraternity that has no peers. Only 560 men and women have gone to space since humans began walking the Earth. Their extraterrestrial journey, they say, has reframed their thinking to transcend the old world order centered on nation-states and race. It certainly did for the Eagles and Soviet cosmonauts who forged lasting friendships together, which endured long after their respective political leaders remained mired in territorial tribalism.

The spirit of Apollo continues with today's astronauts, who view the ISS as a kind of "Shining City" above the Earth, offering the possibility of a new planetary order. "It's just a matter of scale," says Stott. The ISS has been continuously occupied for the past eighteen years by rotating crews from seventeen different nations, who stay alive only through each other's help and the continued upkeep of their life-support systems. Earth is no different, says Stott. "Once we start thinking about Earth as 'Spaceship Earth,' we can adopt those same methods of cooperation and science that keep the space station in orbit."

Will the rest of the world come to see our home planet as the astronauts have? The seeds of an answer are not recent but came instead fifty years ago during that euphoric moment when humans internalized that men really did land on the moon. Something extraordinary happened, remembers Mike Collins, in the way people reacted to the event. It was, he says, the consistent use of the same pronoun.

"After the flight of Apollo 11 the three of us went on an around-the-world trip. Wherever we went, people instead of saying, well *you* Americans did it, everywhere they said *we* did it, *we* humankind, *we* the human race, *we* people did it, and I had never heard of people in different countries use this word *we, we,* as emphatically as we were hearing from Europeans, Asians, Africans. Wherever we went it was 'We finally did it.' I thought that was a wonderful thing, ephemeral, but wonderful."

The collective "*we* humankind" that Collins, Armstrong, and

Aldrin heard never reached the ears of the man who started them on their journey: President Kennedy. Nor did Kennedy see photos of the whole Earth. But he was already sensing its delicate place in the universe, and the collective *we* would show up four times in one of his final speeches before his assassination. For those seeking hope that humanity will one day see itself as a single planetary species, President Kennedy's words transcend the decades with inspiration:

> For, in the final analysis, our most basic common link is that *we* all inhabit this small planet. We all breathe the same air. We all cherish our children's future. And *we* are all mortal.

Acknowledgments

As I began this project in April 2017, Jim Lovell said to me, "Without courage you're not going to do anything, and if you thought that you couldn't write this book, you wouldn't be here." Thank you, Jim, for those words of encouragement.

Few endeavors in life succeed without champions for one's cause. For this book there have been many. I owe particular thanks to Mel Berger, my agent at William Morris Endeavor, who presented the work to Gretchen Young, my editor at Grand Central Publishing. From the outset, both Gretchen and Mel recognized the historical value of this project and the inspiration the Eagles and their wives could provide readers of all ages. They have a superb eye and ear for making complicated subjects easier to read, and that's what they've done here. (Their meticulous assistants, Emily Rosman and David Hinds, helped keep me on track throughout the long journey to publication. I thank you both.) And special callouts to Angelina Krahn for her careful copyedits, production editor Luria Rittenberg for the layout of the book, and Albert Tang for creating such an original jacket cover.

The canon of literature comprising the American space program is considerable, and I owe much to the biographies and autobiographies of all the astronauts who went into space. For readers unfamiliar with the vast number of texts, it is worth singling out a few, which will add to your appreciation of the risks these men and women encountered: Mike Collins's *Carrying the Fire*; William Burrows's *This New Ocean*; James Hansen's biography of Neil Armstrong, *First Man*; Andrew Chaikin's *A Man on the Moon*; Chris Kraft's *Flight*; Gene Kranz's *Failure Is Not an Option*; and Jim Lovell and Jeffrey Kluger's *Apollo 13*.

NASA itself remains the best source of original material documenting America's journey to the moon. Its Oral History Project, in which it systematically interviewed all the key players who made the moon shot possible, is a gift to historians. So are its mission transcripts of all the ground-to-air communications between Mission Control and the astronauts. NASA's collection of photos taken by the astronauts, and onboard film cameras, is a masterwork of magical images of Earth and the moon. Bert Ulrich, NASA's multimedia liaison, along with Connie Moore, the senior photo researcher, each have encyclopedic knowledge of where NASA's best material can be found. NASA is lucky to have such passionate people on staff. I owe them a lot.

Dr. Patricia Santy deserves special mention, since her call for more research on the psychological impact of extended stays in space will do much to ensure that astronauts of the future will be better prepared for their long journeys into our solar system. Many of those young women and men are already being trained through the benefit of scholarships from the Astronaut Scholarship Foundation (ASF). Its team is led by the indefatigable Tammy Sudler, who helped me reach many of the Eagles.

My ultimate champions have been my wife, Marianne, along with my daughters, Lexie and Ariane, who patiently read the first drafts of this work and offered spot-on suggestions. Their love and moral support energized me and made this book possible. Joining them in this effort were some of my dearest friends and armchair historians, Jason Bastis, Jade Wong-Baxter, Andy Cohen, Andy Greenspan, Nancy Hart, John Kirby, and David Timberman. Their advice has been exceptional, as has been the red pen of aerospace historian Bill Burrows, who knows a thing or two about the Apollo program. May we all, one day, travel to the stars together.

BIOGRAPHIES OF THE TWENTY-FOUR EAGLES

Only twelve of the twenty-four Eagles are still living as of this publication. I was fortunate to talk to all but three of them, Dave Scott, Ken Mattingly, and Buzz Aldrin (although I did receive quotes, by e-mail, from Buzz's publicist, Christina Korp). For those who passed away, or were unavailable for personal interviews, I was able to find valuable material from videotaped lectures, particularly at MIT, where many were invited to discuss with students some of the existential questions that have been the focus of this book. The documentary *In the Shadow of the Moon* also proved an excellent source of interview material. NASA has extensive biographical data on each of the twenty-four Eagles, which can be found at: https://www.nasa.gov/astronauts/biographies/former.

For easy reference and the convenience of the reader, below, in alphabetical order, is the complete list of the twenty-four Eagles who went to the moon, with abbreviated versions of their vital information:

EDWIN (BUZZ) EUGENE ALDRIN JR. (Colonel, USAF, Ret.)

Born: January 20, 1930, Montclair, New Jersey
Married: 1st Joan Archer, 2nd Beverly Van Zile, 3rd Lois Driggs Cannon
Children: 3

1951: BS, Mechanical Engineering, US Military Academy, West Point

Pre-NASA: Commissioned as 2nd Lieutenant, US Air Force, as a fighter pilot in Korean War flying 66 combat missions in F-86s. Shot down two MiG-15 aircraft. In 1955 graduated from Squadron Officer School, Maxwell AFB.

1963: ScD, Astronautics, Massachusetts Institute of Technology

1963: Selected as NASA astronaut

NASA career: Gemini 12, Apollo 11 lunar module pilot; second man to walk on the moon

1971: Resigned from NASA

Post-NASA: His life mission is to ensure that America remains at the forefront of space exploration. He is founder and president of ShareSpace Foundation, inspiring young people to pursue interplanetary exploration.

Books: *Magnificent Desolation* (with Ken Abraham), *Mission to Mars, No Dream Is Too High* (with Ken Abraham), *Return to Earth* (with Wayne Warga), *Men from Earth* (with Malcolm McConnell), *Encounter with Tiber* (with John Barnes), *The Return* (with John Barnes), *Reaching for the Moon* (with Wendell Minor), *Look to the Stars*

WILLIAM (BILL) ALISON ANDERS (Major General, USAF Reserve, Ret.)

Born: October 17, 1933, Hong Kong

Married: Valerie Hoard

Children: 6

1955: BS, Electrical Engineering, US Naval Academy

1962: MS, Nuclear Engineering, Air Force Institute of Technology

1979: Advanced Management Program, Harvard Business School

Pre-NASA: US Air Force fighter pilot, Air Defense Command. While at the AF Weapons Laboratory in New Mexico, was responsible for technical management of nuclear power reactor shielding and radiation effects programs.

1964: Selected as NASA astronaut

NASA career: Backup pilot for Gemini 11 and Apollo 11. Lunar module pilot for Apollo 8, the first lunar orbit mission. Famous for the *Earthrise* photo.

1969: Retired from NASA

Post-NASA: Executive secretary, National Aeronautics and Space Council. One of five-member Atomic Energy Commission. Chairman of the newly formed Nuclear Regulatory Commission. US Ambassador to Norway. President and CEO of General Dynamics.

NEIL ALDEN ARMSTRONG (Lieutenant, USN, Ret.)

Born: August 5, 1930, Wapakoneta, Ohio

Died: August 25, 2012 (age 82, following complications from cardiovascular procedures)

Married: 1st Janet Shearon, 2nd Carol Knight

Children: 3

1955: BS, Aeronautical Engineering, Purdue University

1970: MS, Aerospace Engineering, University of Southern California

Pre-NASA: Naval aviator from 1949–52, test pilot at Lewis Flight Propulsion Laboratory, Cleveland, Ohio. Joined National Advisory Committee for Aeronautics (NACA) and was engineer, test pilot, and administrator for NACA and successor agency, NASA.

1962: Selected as NASA astronaut

NASA career: Commander for the Gemini 8 mission. Apollo 11 commander and first man to walk on the moon. Famously said, "That's one small step for man, one giant leap for mankind."

1971: Resigned from NASA

Post-NASA: Taught aerospace engineering at the University of Cincinnati. Deputy associate administrator for Aeronautics, NASA Headquarters,

Washington, DC. Chairman of the Presidential Advisory Committee for the Peace Corps. Spent many years as a consultant and spokesman for several businesses including automotive brand Chrysler.

Books: *First Man* (authorized biography by James R. Hansen)

ALAN LAVERN BEAN (Captain, USN, Ret.)

Born: March 15, 1932, Wheeler, Texas

Died: May 26, 2018 (age 86, sudden illness)

Married: 1st Sue Ragsdale, 2nd Leslie Clem

Children: 2

1955: BS, Aeronautical Engineering, University of Texas

1955: US Naval Test Pilot School

Pre-NASA: 1956–63: Attack Squadron 44 at NAS Jacksonville, Florida, USNTPS, test pilot

1963: Selected as NASA astronaut

NASA career: Served as backup pilot for Gemini 12 and Apollo 9. Was lunar module pilot on Apollo 12. Served as commander on Skylab Mission II (SL-3) and backup spacecraft attendant of American flight crew for the American-Russian Apollo-Soyuz Test Project. Head of the Astronaut Candidate Operations and Training Group as a civilian.

1981: Retired from NASA to focus on his passion for painting

Books: *Alan Bean: Painting Apollo, My Life as an Astronaut, Apollo: An Eyewitness Account* (with Andrew Chaikin)

FRANK FREDERICK BORMAN (Colonel, AF, Ret.)

Born: March 14, 1928, Gary, Indiana

Married: Susan Bugbee

Children: 2

1950: BS, US Military Academy, West Point

1957: MS, Aeronautical Engineering, California Institute of Technology

Pre-NASA: Entered Air Force as a fighter pilot. When selected by NASA, was an instructor at the Aerospace Research Pilot School at Edwards AFB, California.

1962: Selected as NASA astronaut

NASA career: Commander of Gemini 7. In 1967, served on Fire Investigation Board of Apollo 1. Commander of Apollo 8, the first manned lunar orbital mission. Apollo Program resident manager. Field director of NASA's Space Station Task Force.

1970: Retired from NASA

Post-NASA: NASA liaison to President Nixon for Apollo 11. Completed the Harvard Business School's Advanced Management Program. Chief executive officer of Eastern Airlines from 1975–86. Special Presidential Ambassador seeking support for release of American prisoners of war in North Vietnam.

Books: *Countdown: An Autobiography of Frank Borman* (with Robert J. Serling)

EUGENE ANDREW CERNAN (Captain, USN, Ret.)

Born: March 14, 1934, Chicago, Illinois

Died: January 16, 2017 (age 82, natural causes)

Married: 1st Barbara Jean Atchley, 2nd Jan Nanna

Children: 1

1956: BS, Electrical Engineering, Purdue University

1963: MS, Aeronautical Engineering, US Naval Postgraduate School

1963: Selected as NASA astronaut

NASA career: Pilot for Gemini 9. Lunar module pilot on Apollo 10 and the commander for Apollo 17 in 1972 on the last manned mission to the moon. He was the last man to walk on the moon.

1976: Retired from the Navy and NASA.

Post-NASA: Special assistant to the program manager of Project Apollo at the Johnson Space Center. He went into private business and was a contributor to ABC News. His company, the Cernan Corporation, manages and consults on energy and aerospace.

Books: *The Last Man on the Moon* (with Don Davis)

MICHAEL COLLINS (Major General, USAF, Ret.)

Born: October 31, 1930, Rome, Italy

Married: Patricia Finnegan

Children: 3

1952: BS, US Military Academy, West Point

Pre-NASA: Air Force flight training and then served as an experimental flight test officer at Edwards AFB, California. Entered USAF Research Pilot School.

1963: Selected as NASA astronaut

NASA career: Backup pilot for Gemini 7, pilot on Gemini 10, command module pilot on Apollo 11

1970: Retired from NASA

Post-NASA: Assistant Secretary of State for Public Affairs, 1970. Director of the National Air and Space Museum, Smithsonian Institution, Washington, DC, 1971–78. Advanced Management Program, Harvard Business School, 1974.

Books: *Carrying the Fire, Flying to the Moon: An Astronaut's Story, Liftoff: The Story of America's Adventure in Space*

CHARLES (PETE) CONRAD JR. (Captain, USN, Ret.)

Born: June 2, 1930, Philadelphia, Pennsylvania

Died: July 8, 1999 (age 69 from complications following a motorcycle accident)

Married: 1st Jane DuBose, 2nd Nancy Crane

Children: 4

1953: BS, Aeronautical Engineering, Princeton University

1957: Entered US Naval Test Pilot School

1962: Selected as NASA astronaut

NASA career: Served as the commander on Gemini 11 mission and on Apollo 12, the second lunar landing. Was commander on Skylab II, the first manned voyage to Skylab, which was the first space station.

1973: Retired from Navy and NASA

Post-NASA: Worked in private sector with ATCC and McDonnell Douglas. Developed large-scale projects for making commercial spaceflight a practical reality.

Books: *Rocketman* (authorized biography by Nancy Conrad and Howard Klausner)

CHARLES (CHARLIE) MOSS DUKE JR.
(Brigadier General, USAF, Ret.)

Born: October 3, 1935, Charlotte, North Carolina

Married: Dorothy Meade Claiborne

Children: 2

1957: BS, Naval Sciences, US Naval Academy

1964: MS, Aeronautics, Massachusetts Institute of Technology

1965: Graduated from the Air Force Aerospace Research Pilot School

Pre-NASA: Finished advanced flight training, served as fighter interceptor pilot at Ramstein Air Base, Germany, and instructor at the Air Force Aerospace Research Pilot School

1966: Selected as NASA astronaut

NASA career: Support crew for Apollo 10, CAPCOM for Apollo 11, backup lunar module pilot for Apollo 13 and 17, lunar module pilot for Apollo 16, and tenth man to walk on the moon

1975: Retired from NASA

Post-NASA: Started his own business, Duke Investments, president of Duke Ministry for Christ

Books: *Moonwalker* (with Dotty Duke)

RONALD (RON) ELLWIN EVANS (Captain, USN, Ret.)

Born: November 10, 1933, St. Francis, Kansas

Died: April 7, 1990 (age 56, heart attack)

Married: Janet Merle

Children: 2

1956: BS, Electrical Engineering, University of Kansas

1957: Completed flight training through Navy ROTC, University of Kansas

1961–62: Combat flight instructor with VF-124

1964: MS, Aeronautical Engineering, US Naval Postgraduate School

Pre-NASA: Served in Vietnam (1964–66)

1966: Selected as NASA astronaut

NASA career: Command module pilot for Apollo 17, served in development of
NASA's space shuttle program

1977: Resigned from NASA

Post-NASA: Executive in the coal industry

Books: *Apollo and America's Moon Landing Program* (with Gene Cernan and
Harrison Schmitt)

RICHARD FRANCIS GORDON JR. (Captain, USN, Ret.)

Born: October 5, 1929, Seattle, Washington

Died: November 6, 2017 (age 88, cancer)

Married: 1st Barbara Field, 2nd Linda Saunders

Children: 6

1951: BS, Chemistry, University of Washington

1953: Received his naval aviator wings and then attended flight school and
studied jet transitional training

1957: Enrolled in US Naval Test Pilot School, Maryland, and served as a test
pilot until 1960

1963: Selected as NASA astronaut

NASA career: Pilot for Gemini 11, command module pilot for Apollo 12

1972: Retired from NASA

Post-NASA: Executive vice president of the New Orleans Saints professional football club, other executive positions in technology and engineering industries

FRED WALLACE HAISE JR. (Captain, USAF, Ret.)

Born: November 14, 1933, Biloxi, Mississippi

Married: 1st Mary Grant, 2nd F. Patt Price

Children: 4

1959: BS, Aeronautical Engineering, University of Oklahoma

Pre-NASA: Research pilot at NASA Flight Research Center, Air Force tactical fighter pilot, flight instructor, US Navy Advanced Training Command, US Marine Corps fighter pilot

1966: Selected as NASA astronaut

NASA career: Lunar module pilot for Apollo 13, technical assistant to manager of Space Shuttle Orbiter Project, commander for space shuttle approach and landing test flights

1979: Retired from NASA

Post-NASA: Vice president of Space Programs, Grumman Aerospace Corporation, president of Grumman Technical Services

JAMES (JIM) BENSON IRWIN (Colonel, USAF, Ret.)

Born: March 17, 1930, Pittsburgh, Pennsylvania

Died: August 8, 1991 (age 61, heart attack)

Married: Mary Ellen Monroe

Children: 5

1951: BS, Naval Science, US Naval Academy

1957: MS, Aeronautical Engineering and Instrumentation Engineering, University of Michigan

1961: Graduated from the Air Force Experimental Test Pilot School

1963: Graduated from the Air Force Aerospace Research Pilot School

Pre-NASA: Commissioned in the Air Force, flight training, served as chief of the Advanced Requirements Branch at Headquarters Air Defense Command

1966: Selected as NASA astronaut

NASA career: Eighth man to walk on the moon, Apollo 15

1972: Resigned from NASA

Post-NASA: Founder and chairman of the board for the High Flight Foundation, Colorado Springs, Colorado

Books: *More Than Earthlings, To Rule the Night* (with William A. Emerson Jr.), *More Than an Ark on Ararat* (with Monte Unger)

JAMES (JIM) ARTHUR LOVELL JR. (Captain, USN, Ret.)

Born: March 25, 1928, Cleveland, Ohio

Married: Marilyn Gerlach

Children: 4

1952: BS, US Naval Academy

1958: US Naval Test Pilot School, NATC, Patuxent River, Maryland

1961: Aviation Safety School, University of Southern California

Pre-NASA: Safety engineer, Fighter Squadron 101, NAS, Oceana, Virginia

1962: Selected as NASA astronaut

NASA career: Gemini 7 and 12, Apollo 8, and commander of Apollo 13

1967: Appointed by President Johnson as consultant for Physical Fitness and Sports

1970: Became the chairman of the Physical Fitness Council under President Nixon

1971: Advanced Management Program, Harvard Business School

1973: Retired from NASA

Post-NASA: Became president and chief executive of Bay-Houston Towing
Company, Houston, Texas; president of Fisk Telephone Systems, Inc.; vice
president of Business Communications Systems

Books: *Lost Moon: The Perilous Voyage of Apollo 13* (with Jeffrey Kluger)

THOMAS KENNETH MATTINGLY II (Rear Admiral, USN, Ret.)

Born: March 17, 1936, Chicago, Illinois

Married: Elizabeth Bailey

Children: 1

1958: BS, Aeronautical Engineering, Auburn University

1958: Ensign in the Navy, received his wings in 1960 and then became a student
in the AF Aerospace Research Pilot School

1966: Selected as NASA astronaut

NASA career: Astronaut representative in development and testing of Apollo
spacesuit and backpack; command module pilot for Apollo 13 but removed
72 hours prior due to measles exposure; command module pilot for Apollo
16; head of astronaut office support to shuttle transport system program;
assisted with Orbital Flight Test Program; backup commander for STS-2,
3, 4, 5; head of Astronaut Office DOD Support Group

1985: Retired from NASA

Post-NASA: Director for Grumman's Space Station Support Division; head of
Atlas Booster Program, General Dynamics, San Diego, California; vice
president at Lockheed Martin for X-33 development program

Books: *Apollo and America's Moon Landing Program*

EDGAR (ED) DEAN MITCHELL (Captain, USN, Ret.)

Born: September 17, 1930, Hereford, Texas

Died: February 4, 2016 (age 85 after short illness)

Married: 1st Louise Randall, 2nd Anita Rettig, 3rd Sheilah Ledbetter

Appendix

Children: 6 (3 adopted)

1952: BS, Industrial Management, Carnegie Mellon University

1961: BS, Aeronautics, US Naval Postgraduate School

1964: ScD, Aeronautics and Astronautics, Massachusetts Institute of Technology

Pre-NASA: Manned Spacecraft Center, Houston, Texas; pilot in the Navy; test pilot certification, Air Force Aerospace Research Pilot School

1966: Selected as NASA astronaut

NASA career: Sixth man to walk on the moon on Apollo 14

1972: Retired from NASA

Post-NASA: Founded Institute of Noetic Sciences in 1973, a foundation dedicated to research on the nature of consciousness

Books: *The Way of the Explorer* (with Dwight Williams), *Earthrise: My Adventures as an Apollo 14 Astronaut*, *Psychic Exploration*

STUART "SMOKY" ALLEN ROOSA (Colonel, USAF, Ret.)

Born: August 16, 1933, Durango, Colorado

Died: December 12, 1994 (age 61, pancreatitis)

Married: Joan Barrett

Children: 4

1960: BS, Aeronautical Engineering, University of Colorado

Pre-NASA: Aerospace Research Pilot School, joined the Air Force, attended Gunnery School at Del Rio and Luke AFB, graduate of Aviation Cadet Program, experimental test pilot, Edwards AFB, California, fighter pilot at Langley AFB, Virginia

1966: Selected as NASA astronaut

NASA career: Command module pilot, Apollo 14; backup command module for Apollo 16 and 17; space shuttle program

1973: Advanced Management Program, Harvard Business School

1976: Retired from NASA

Post-NASA: Executive corporate positions at US Industries, Inc., USI Middle East Development, Ltd., Charles Kenneth Campbell Investments, and Gulf Coast Coors, Inc.

HARRISON H. SCHMITT

Born: July 3, 1935, Santa Rita, New Mexico

Married: Teresa Fitzgibbon

Children: None

1957: BS, Science, California Institute of Technology

1957–1958: University of Oslo

1964: PhD, Geology, Harvard University

Pre-NASA: Teaching fellow at Harvard, accomplished geologist at US Geological Survey's Astrogeology Center instructing astronauts during their geological field trips

1965: Selected as NASA scientist-astronaut

NASA career: First sent to 53-week flight training course at Williams AFB, Arizona; served as backup lunar module pilot for Apollo 15 and lunar module pilot for Apollo 17; chief of Scientist-Astronauts and NASA assistant administrator for Energy Programs

1975: Resigned from NASA

Post-NASA: Elected as US Senator in New Mexico; worked as a professor, consultant, writer, and speaker

Books: *Return to the Moon*

DAVID RANDOLPH SCOTT (Colonel, USAF, Ret.)

Born: June 6, 1932, San Antonio, Texas

Married: 1st Ann Lurton Ott, 2nd Margaret Black

Children: 2

1954: BS, US Military Academy, West Point

1962: MS, Aeronautics Engineering and Astronautics, Massachusetts Institute of Technology

Pre-NASA: Completed pilot training at Webb AFB, then gunnery training at Laughlin and Luke AFBs; graduated from AF Test Pilot School

1963: Selected as NASA astronaut

NASA career: Gemini 8; command module pilot, Apollo 9; commander, Apollo 15, and seventh man to walk on the moon; one of only three to have flown earth orbital and lunar Apollo missions. NASA Director, Dryden Flight Research Center.

1975: Retired from Air Force

1977: Retired from NASA

Post-NASA: President of Scott Science and Technology, Inc.

Books: *Two Sides of the Moon* (with Alexei Leonov)

ALAN BARTLETT SHEPARD JR. (Rear Admiral, USN, Ret.)

Born: November 18, 1923, East Derry, New Hampshire

Died: July 21, 1998 (age 74, leukemia)

Married: Louise Brewer

Children: 3

1944: BS, US Naval Academy

1951: Graduated from the Naval Test Pilot School

1957: Naval War College, Newport, Rhode Island

1959: Selected as NASA astronaut

NASA career: Freedom 7 Mercury; first American in space; chief of the Astronaut Office; commander on Apollo 14, and fifth man to walk on the moon

1974: Retired from NASA

Post-NASA: Chairman of Marathon Construction Corporation, founder of Seven Fourteen Enterprises, and chairman of Mercury Seven Foundation

Books: *Moon Shot: The Inside Story of America's Apollo Moon Landings* (with Deke Slayton and Jay Barbree)

THOMAS PATTEN STAFFORD (Lieutenant General, USAF, Ret.)

Born: September 17, 1930, Weatherford, Oklahoma

Married: 1st Faye Shoemaker, 2nd Linda Ann Dishman

Children: 4

1952: BS, US Naval Academy

Pre-NASA: Selected by lottery to join the US Air Force

1958: Attended US Air Force Test Pilot School, Edwards AFB, USAF Experimental Test Pilot School

1962: Selected as NASA astronaut

NASA career: Pilot for Gemini 6, commander of Gemini 9, commander of Apollo 10, deputy director of Flight Crew Operations, commander of the Apollo-Soyuz Test Project flight

1979: Retired from NASA

Post-NASA: Commander of the Air Force Flight Test Center, Edwards AFB; chair of numerous NASA committees; consultant and adviser to other corporate boards

Books: *We Have Capture* (with Michael Cassutt)

JOHN (JACK) LEONARD SWIGERT JR. (Captain, USAF, Ret.)

Born: August 30, 1931, Denver, Colorado

Died: December 27, 1982 (age 51, bone cancer)

Married: Never married

Children: None

1953: BS, Mechanical Engineering, University of Colorado

1965: MS, Aerospace Science, Rensselaer Polytechnic Institute

Pre-NASA: Air Force (1953–56), fighter pilot in Japan and Korea, Massachusetts Air National Guard (1957–60), Connecticut Air National Guard (1960–65), North American Aviation engineering test pilot, Pratt and Whitney engineering test pilot

1966: Selected as NASA astronaut

NASA career: Command module pilot, Apollo 13

1973: Took leave of absence from NASA and became executive director of what later became known as the Committee on Science and Technology, US House of Representatives

1977: Resigned from NASA

Post-NASA: Full-time politician; vice president, BDM Corporation (1979); elected to House of Representatives in 1982 but died before he was sworn in

ALFRED (AL) MERRILL WORDEN (Colonel, USAF, Ret.)

Born: February 7, 1932, Jackson, Michigan

Married: 1st Pamela Vander Beek, 2nd Jill Lee Hotchkiss

Children: 3

1955: BS, Military Science, US Military Academy, West Point

1963: MS, Astronautical and Aeronautical Engineering and Instrumentation Engineering, University of Michigan

1965: Empire Test Pilots' School, England

Pre-NASA: Air Force with flight training at Moore Air Base, Laredo AFB, and Tyndall AFB

1966: Selected as NASA astronaut

NASA career: Astronaut support crew, Apollo 9; backup command module pilot, Apollo 12; command module pilot, Apollo 15; senior aerospace scientist, NASA Ames Research Center (1972–73); chief, Systems Study Division, Ames (1973–75)

1975: Retired from NASA

Post-NASA: President of Maris Worden Aerospace, Inc.; staff vice president of BF Goodrich Aerospace; director at Astronaut Scholarship Foundation

Books: *Falling to Earth, Hello Earth: Greetings from Endeavour, I Want to Know about a Flight to the Moon* (with Fred Rogers)

JOHN WATTS YOUNG (Captain, USN, Ret.)

Born: September 24, 1930, San Francisco, California

Died: January 5, 2018 (age 87, pneumonia)

Married: 1st Barbara White, 2nd Susy Feldman

Children: 2

1952: BS, Aeronautical Engineering, Georgia Institute of Technology

Pre-NASA: Enlisted in Navy, flew fighter planes for 4 years; served as a test pilot at Navy's Air Test Center for 3 years

1962: Selected as NASA astronaut

NASA career: Gemini 3, commander of Gemini 10, command module pilot of Apollo 10, commander of Apollo 16, and ninth person to walk on the moon; chief of Space Shuttle Branch at the Johnson Space Center; commander of space shuttle missions during work on STS-1, STS-9

Post-NASA: Worked for NASA for 42 years, retiring at age 74, the longest of any of the astronauts

Books: *Forever Young* (with James R. Hansen)

Notes

INTRODUCTION

Page 1   **"I was a lot more worried:** David Sington, *In the Shadow of the Moon* (2007; New York, ThinkFilm, 2008), DVD.

Page 2   **"I remember when Frank Borman:** *Meet the Press, July 18, 1999.* Directed by Leigh Sutherland. Produced by Betsy Fischer Martin. NBCUniversal, 1999, audio transcript, 36:50, https://search.alexanderstreet.com/view/work/bibliographic_entity%7Cvideo_work%7C2479055.

Page 2   **Fate has ordained that:** *In Event of Moon Disaster,* https://www.archives.gov/files/presidential-libraries/events/centennials/nixon/images/exhibit/rn100-6-1-2.pdf.

Page 3   **"Pete looks over at me:** Sington, *Shadow of the Moon.*

Page 4   **"The unknowns were rampant,":** James R. Hansen, *First Man: The Life of Neil Armstrong* (Simon & Schuster, New York, 2005), 529, print version.

Page 6   **pursuit of what he called *the common good*:** Carnes Lord, *Aristotle's Politics,* 2nd ed. (The University of Chicago Press, Chicago and London, 2013), 1.

Page 7   **Like that of Ulysses:** Garik Israelian and Brian May, *Starmus: 50 Years of Man in Space* (Starmus, 2014), 17.

Chapter One: THE REAL RIGHT STUFF: SELECTING THE EAGLES

Page 11   **"Test pilot experience was critical,":** Sington, *Shadow of the Moon.*

Page 12   **"That group of astronauts:** Ibid.

Page 12   **"I thought I had the best job:** Ibid.

Page 12   **"It certainly sounded very challenging,":** Ibid.

Page 13   **"We had to endure:** Frank Borman with Robert J. Serling, *Countdown: An Autobiography* (Silver Arrow Books, New York, 1988), 88.

Notes

Page 14   "They gave us inkblots to identify: John W. Young with James R. Hansen, *Forever Young: A Life of Adventure in Air and Space* (University Press of Florida, Gainesville, 2012), 212, e-book.

Page 16   Candidates should have a high level: Patricia A. Santy, *Choosing the Right Stuff: The Psychological Selection of Astronauts and Cosmonauts* (Praeger Publishers, Westport, CT, 1994), 11.

Page 17   "very few fit the popular concept: Ibid, 21.

Page 18   Because of the possibility: Ibid.

Page 20   "I'm sorry, Doc. I can't.": Nancy Conrad and Howard A. Klausner, *Rocketman: Astronaut Pete Conrad's Incredible Ride to the Moon and Beyond* (New American Library, New York, 2005), 324, e-book.

Page 21   "Oh, a vagina. That's definitely: Ibid.

Page 21   "a conviction that: Santy, *Choosing the Right Stuff*, 21.

Page 21   "It's a very methodical, analytical: Douglas MacKinnon and Joseph Baldanza, *Footprints: The 12 Men Who Walked on the Moon Reflect on Their Flights, Their Lives, and the Future* (Acropolis Books, Washington, DC, 1989), 18.

Page 22   "I thought he'd pulled a showboat: Borman with Serling, *Countdown*, 86.

Page 24   "I lived my early years in Depression-hit: Young with Hansen, *Forever Young*, 201, e-book.

Page 25   "I developed maturity: Borman with Serling, *Countdown*, 18.

Page 26   "The aim of every political constitution: *The Federalist Papers*, no. 57, https://en.wikiquote.org/wiki/The_Federalist.

Chapter Two: THE TECHNIQUES FOR CONQUERING FEAR

Page 33   "I always thought of myself as one of: Sington, *Shadow of the Moon*.

Page 37   "Houston, Tranquility base here.: Apollo 11 Technical Air-to-Ground Voice Transcription, July 1969, 317, https://www.jsc.nasa.gov/history/mission_trans/AS11_TEC.PDF.

Page 37   "Roger, Tranquility.: Ibid.

Page 38   "You deal with what's in front of you: Sington, *Shadow of the Moon*.

Page 39   "The worst thing is seeing the flak,: Lord Moran, *The Anatomy of Courage: The Classic WWI Study of the Psychological Effects of War* (Constable and Company, 1945), 312, e-book.

Notes

Page 40   **"I believe we've had a problem:** Apollo 13 Technical Air-to-Ground
           Voice Transcription, April 1970, 160, https://www.jsc.nasa.gov
           /history/mission_trans/AS13_TEC.PDF.

Page 40   **"This is Houston. Say again, please,":** Ibid.

Page 40   **"Houston, we've had a problem.":** Ibid.

Page 41   **"The crew conflict you saw:** Lianne Hart, "Bit of Hollywood Goes
           a Long Way for NASA: Movies: The Box-Office Success of
           'Apollo 13' Renews Interest in Both That Flight and the Space
           Program," *Los Angeles Times*, July 18, 1995.

Page 47   **"This is the young man,:** Richard Nixon, "334—Remarks at a Din-
           ner in Los Angeles Honoring the Apollo 11 Astronauts," Plaza
           Hotel, August 13, 1969, transcript, American Presidency Proj-
           ect, http://www.presidency.ucsb.edu/ws/?pid=2202.

Page 48   **"Okay, we just lost the platform,:** Apollo 12 Technical Air-to-Ground
           Voice Transcription, November 1969, 1, https://www.jsc.nasa.gov
           /history/mission_trans/AS12_TEC.PDF.

Page 49   **"What the hell is SCE?:** Nancy Conrad and Howard A. Klausner,
           *Rocketman: Astronaut Pete Conrad's Incredible Ride to the Moon
           and Beyond*, 464.

Page 50   **"We had been trained:** David Fairhead, *Mission Control: The
           Unsung Heroes of Apollo* (2017, Gravitas Ventures), DVD.

Chapter Three: THE WIVES: BRAVER THAN THE EAGLES

Page 56   **"in the year 1952 alone, sixty-two pilots:** James R. Hansen, *First
           Man: The Life of Neil Armstrong* (Simon & Schuster, New York,
           2005), 986, electronic version.

Page 58   **"We take an additional risk:** President John F. Kennedy, "Special
           Message to the Congress on Urgent National Needs," deliv-
           ered in person before a joint session of Congress, May 25, 1961,
           partial transcript, https://www.nasa.gov/vision/space/features
           /jfk_speech_text.html.

Page 59   **"We go into space because:** Ibid.

Page 60   **"Because I had been:** Fairhead, *Mission Control*.

Page 60   **"Who am I, Susan?:** Borman with Serling, *Countdown*, 295.

Page 61   **"the astronaut's astronaut,:** Ibid., 294.

Page 70   **"I sat next to Dad:** Borman with Serling, *Countdown*, 14.

Page 72   **It is said that domestic troubles:** Nancy Shea, *The Air Force Wife* (Harper & Brothers, 1951), 1.

Page 73   **Kraft had grown to respect:** Borman with Serling, *Countdown*, 198.

Page 73   **"Just remember, Mom:** Ibid.

Page 74   **"They had people looking into:** "NASA Wives and Families," PBS, American Experience, October 31, 2005, https://www.pbs.org/wgbh /americanexperience/features/moon-nasa-wives-and-families/.

Page 78   **"I have memories of running:** *PBS, American Experience: Race to the Moon,* October 6, 2005, http://www.shoppbs.pbs.org/wgbh /amex/moon/sfeature/sf_families.html.

Page 79   **"I was in second grade.:** Ibid.

Chapter Four: LEADERSHIP LESSONS AND DOING THE IMPOSSIBLE

Page 95   **"I was watching TV at home:** Christopher Novak, *Conquering Adversity: Six Strategies to Move You and Your Team through Tough Times* (Cornerstone Leadership Institute, Dallas, 2004), https://bit.ly/2PRKq8z.

Page 97   **"Had that not occurred these men:** Richard Nixon, "122—Remarks on Presenting the Presidential Medal of Freedom to Apollo 13 Mission Operations Team in Houston," Manned Space Center, April 18, 1970, transcript, American Presidency Project, http:// www.presidency.ucsb.edu/ws/?pid=2471.

Page 99   **"We would like to give a special thanks:** Hansen: *First Man*, 3313, electronic version.

Page 101   **"an ambitious, exploratory and:** WhatIs.com, s.v. "Moonshot," by Margaret Rouse, https://whatis.techtarget.com/definition /moonshot.

Page 101   **"Was Kennedy a visionary,:** Sington, *Shadow of the Moon.*

Page 101   **We choose to go:** John F. Kennedy Moon Speech at Rice Stadium, https://er.jsc.nasa.gov/seh/ricetalk.htm.

Page 102   **"Do what? Moon!:** Sington, *Shadow of the Moon.*

Page 104   **"Politically it was about beating:** Ibid.

Page 104   **"We knew what had to be done,":** Launius, "Managing the Moon," 12.

Page 104 **"I told him that I thought:** Ibid.

Page 109 **"Nobody really knew how:** Sington, *Shadow of the Moon.*

Page 109 **"I looked for people:** Launius, "Managing the Moon," 17.

Page 110 **"Another thing that I think was extraordinary,:** Ibid., 18.

Page 110 **"I mean, NASA responsibilities:** Ibid.

Page 110 **"I think we had a group that:** Ibid.

Page 110 **"And the other thing you have to realize,":** Ibid.

Page 111 **"We grew up telling each other:** Ibid.

Page 112 **You had to set up the interfaces:** Ibid., 16.

Page 113 **"What we did in the early days,":** Sington, *Shadow of the Moon.*

Page 114 **"It isn't that we don't trust:** Charles Murray and Catherine Bly Cox, *Apollo: The Race to the Moon* (Simon & Schuster, New York, 1989), 184.

Page 115 **"The animosity between my people:** Chris Kraft, *Flight: My Life in Mission Control* (Penguin Group, New York, 2001), 251.

Page 116 **"He was unquestionably a brilliant:** Borman with Serling, *Countdown*, 178.

Page 117 **From this day forward, Flight Control:** Gene Kranz, *Failure Is Not an Option: Mission Control from Mercury to Apollo 13 and Beyond* (Simon & Schuster, New York, 2000), 1058, electronic version.

Page 118 **"When we had the fire,":** Launius, "Managing the Moon," 27.

Page 118 **"As a result of the accident,:** Ibid.

Page 119 **"He had a Configuration Control Board:** Ibid., 30.

Page 119 **"I got into the Apollo simulator:** Borman with Serling, *Countdown*, 187.

Page 120 **"There is a time to be conservative:** Launius, "Managing the Moon," 26.

Page 120 **"the single greatest:** Alan Shepard and Deke Slayton with Jay Barbree, *Moon Shot: The Inside Story of America's Apollo Moon Landings* (Turner Publishing Inc, Atlanta, 1994).

Page 120 **"We hear from the CIA that:** Sington, *Shadow of the Moon.*

Page 120 **"If they orbit the moon before:** Ibid.

Page 121 **"We changed our plans:** Ibid.

Page 121 **"We spent a great deal of time,:** Launius, "Managing the Moon," 26.

Page 121 **"We knew what we were doing,":** Ibid., 31.

Chapter Five: CHANGED BY A VIEW: THE COSMIC LIGHTHOUSE

Page 127   **This is Apollo 8, coming to you:** Apollo 8 Technical Air-to-Ground Voice Transcription, December 1968, tape 57, 5, https://www .jsc.nasa.gov/history/mission_trans/AS08_TEC.PDF.

Page 127   **Well, Frank, my thoughts are very:** Ibid., tape 57, 6.

Page 132   **"When the sun is shining on the surface:** Sington, *Shadow of the Moon*.

Page 133   **"When you land on the moon,:** Ibid.

Page 133   **"I think the feeling I had the whole:** Ibid.

Page 134   **"We were looking at things:** Ibid.

Page 136   **"What I keep imagining is if I am some:** Apollo 8 Technical Air-to-Ground Voice Transcription, tape 37, 10.

Page 136   **"Don't see anybody waving;:** Ibid., tape 37, 11.

Page 136   **And seeing it so, one question:** Archibald MacLeish, "A Reflection: Riders on Earth Together, Brothers in Eternal Cold," *New York Times*, December 25, 1968.

Page 138   **You develop an instant global:** Quoted in Edgar G. Mitchell, International Space Hall of Fame at the New Mexico Museum of Space History, http://www.nmspacemuseum.org/halloffame /detail.php?id=45.

Page 138   **"I really believe that if the political leaders:** Michael Collins, *Carrying the Fire: An Astronaut's Journeys* (Farrar, Straus and Giroux, New York, 1974), 470.

Page 139   **"We went to the Moon as technicians;:** Mitchell, Space Hall of Fame.

Page 140   **"20 million Americans took to the streets,:** "The History of Earth Day," Earth Day Network, https://www.earthday.org /about/the-history-of-earth-day.

Page 140   **A unique day in American history:** CBS News Special: Earth Day, April 22, 1970, https://www.cbsnews.com/news/almanac-the -first-earth-day/.

Page 141   **"The living Earth complains of fever,":** James E. Lovelock, *The Revenge of Gaia* (Basic Books, New York, 2006), 1.

Page 141   **"The dead zone in the Gulf of:** Collins, *Carrying the Fire* (2009), xviii.

Page 142 . "The earth, our home, is beginning: Pope Francis, *Encyclical Letter Laudato Si' of the Holy Father Francis on Care for Our Common Home,* 17, https://w2.vatican.va/content/dam/francesco/pdf /encyclicals/documents/papa-francesco_20150524_enciclica -laudato-si_en.pdf.

Page 142 Never have we so hurt and mistreated: Ibid., 39.

Page 143 If somebody'd said before the flight: Andrew Chaikin with Victoria Kohl, *Voices from the Moon: Apollo Astronauts Describe Their Lunar Experiences* (Viking Studio, New York, 2009), 99.

Page 143 "The most wonderful thing I saw was the Earth.: MIT AeroAstro Video, April 27, 2017, http://aeroastro.mit.edu/videos/apollo-15s -worden-and-mission-planner-el-baz-visit-aeroastro-4-27-2017.

Chapter Six: THE EAGLES AND THE GOD QUESTION

Page 147 "When I looked back and saw that: Ron Judd, "With a View from Beyond the Moon, an Astronaut Talks Religion, Politics and Possibilities," *Seattle Times,* December 7, 2012.

Page 148 "We are challenged both mystically and: Eugene Kennedy, "Earth-rise," *New York Times,* April 15, 1979, https://www.nytimes .com/1979/04/15/archives/earthrise-the-dawning-of-a-new -spiritual-awareness.html.

Page 149 "Suddenly I realized that the molecules: Sington, *Shadow of the Moon.*

Page 150 Fast-forward to the: Pew Research Center, April 25, 2018, http:// www.pewresearch.org/fact-tank/2018/04/25/key-findings -about-americans-belief-in-god.

Page 153 "I looked and looked but I didn't see: Paul Schlieker, *What's Missing Inside You? Discover the Leading Character in the Story of Your Life* (Xulon Press, 2006), 17.

Page 154 "I think it would be inappropriate,": Robert Zimmerman, *Genesis: The Story of Apollo 8* (Dell Publishing, New York, 1998), 231.

Page 155 "You're in the wrong part: Billy Watkins, *Apollo Moon Missions: The Unsung Heroes* (Praeger, Westport, CT, 2005), 70.

Page 155 "Everything I wrote," said Laitin,: Ibid., 70.

Page 156   **We are now approaching the lunar:** Apollo 8 Technical Air-to-Ground Voice Transcription, tape 58, 3.

Page 156   **"I was enraptured, transported:** Kranz, *Failure*, 1267.

Page 159   **"I would like to request a few moments:** Buzz Aldrin, *Magnificent Desolation: The Long Journey Home from the Moon* (Harmony Books, New York, 2009), 26.

Page 159   **"I poured the wine:** *The Moon Communion of Buzz Aldrin That NASA Didn't Want to Broadcast*, Huff Post, July 19, 2014, updated Dec 06, 2017, https://www.huffingtonpost.com/2014/07/19/moon -communion-buzz-aldrin_n_5600648.html.

Page 159   **"I don't really know what that has:** Hansen: *First Man*, 34.

Page 160   **The application asked him to list:** Ibid., 33.

Page 161   **The most beautiful and most profound emotion:** Lincoln Barnett, *The Universe and Dr. Einstein* (William Morrow, New York, 1948), 108.

Page 164   **"Religion is man-made.:** Sington, *Shadow of the Moon.*

Chapter Seven: PARADISE FOUND: AND ON THE BRINK

Page 171   **"Standing on the Moon looking back:** Steven Ferrey, *Environmental Law: Examples & Explanations* (Wolters Kluwer, New York, 2016). Chapter 6.

Page 173   **"I have not complained about the weather:** Sington, *Shadow of the Moon.*

Page 173   **"I can remember the beautiful:** Ibid.

Page 174   **"Hopefully, by getting a little farther:** Hansen, *First Man*, 3388, electronic version.

Page 174   **We travel together, passengers:** Speech to the Economic and Social Council of the United Nations, Geneva, Switzerland, July 9, 1965.

Page 176   **While we send men to the Moon,:** Jesse Jackson, "Reactions to Man's Landing on the Moon Show Broad Variations in Opinions," *New York Times*, July 21, 1969, 7.

Page 177   **"How can you spend all this money:** Borman with Serling, *Countdown*, 235.

Page 177 **"We've wasted all this money,":** David Scott and Alexei Leonov, *Two Sides of the Moon: Our Story of the Cold War Space Race* (Thomas Dunne Books, St. Martin's Griffin, New York, 2013), 241.

Page 181 **"Earth has changed a lot since we:** Sington, *Shadow of the Moon.*

Page 181 **"It truly is an oasis,":** Ibid.

Page 182 **"We humans are responsible for:** Ibid.

Page 183 **"What is the prime imperative:** "Apollo 15's Worden and Mission Planner El Baz Visit AeroAstro 4-27-2017," MIT Aeronautics and Astronautics Department, April 27, 2017, video, 1:54:03, http://aeroastro.mit.edu/videos/apollo-15s-worden-and -mission-planner-el-baz-visit-aeroastro-4-27-2017.

Chapter Eight: **GRADATIM FEROCITER: THE FUTURE OF SPACE TRAVEL**

Page 190 **"You have got to live in something,":** Catherine Clifford, "Buzz Aldrin Says This Is the Problem with Elon Musk's Plans for Mars," CNBC, March 15, 2017, https://www.cnbc.com/2017/03/15/buzz -aldrin-says-this-is-the-problem-with-elon-musks-mars-plan .html.

Page 190 **"Our Stratolaunch carrier:** Paul G. Allen, *Tackling the Space Challenge,* June 20, 2016, https://linkedin.com/pulse/tackling-space -challenge-paul-g-allen/.

Page 191 **"I often tell people that my middle:** Buzz Aldrin, Buzz Aldrin Space Innovation Gala at Kennedy Space Center, Florida, July 15, 2017.

Page 197 **"Our space program has been:** Sophie Shevardnadze, "U.S. Space Program Is Dead-End," RT News, July 27, 2011, https://www .rt.com/news/space-program-usa-anders/.

Page 202 **"If you're building a flying vehicle,:** Alan Boyle, "Gradatim Ferociter! Jeff Bezos Explains Blue Origin's Motto, Logo...and the Boots," Geek Wire, October 24, 2016, https://www.geekwire .com/2016/jeff-bezos-blue-origin-motto-logo-boots/.

Page 202 **"It's just a symbol of the:** Ibid.

Page 203   **"A lot of people [wondered] what's:** William Harwood, "SpaceX's Tesla's Out-of-This-World View," CBS News, February 26, 2018, https://www.cbsnews.com/news/spacex-tesla-launched-on-falcon-heavy-out-of-this-world-view/.

Page 204   **"With climate change, overdue:** Julia Zorthian, "Stephen Hawking Says Humans Have 100 Years to Move to Another Planet," *Time*, May 4, 2017, http://time.com/4767595/stephen-hawking-100-years-new-planet/.

Page 207   **"I'd like to see continuity in this:** Brian Cox, Q&A Panel, Astronaut Scholarship Foundation Gala, Kennedy Space Center, Florida, July 21, 2018.

Page 207   **"It's all hot air until someone:** Nola Taylor Redd, "Trump's 'Back to the Moon' Directive Leaves Some Scientists with Mixed Feelings," Space.com, December 28, 2017, https://space.com/39221-trump-moon-space-policy-planetary-scientists.html.

Page 213   **"Based on our record here on Earth:** Israelian and May, *Starmus*, 17.

Chapter Nine: THE NOBLEST JOURNEY OF ALL

Page 219   **Because of what you have done:** Richard Nixon, "Telephone Conversation with the Apollo 11 Astronauts on the Moon," July 20, 1969 (P-690714 7/20/1969 1:14), https://nixonlibrary.gov/forkids/speechesforkids/moonlanding/moonlandingcall.pdf.

Page 220   **"I'm impressed with the margin:** William Stockton, "Anders Says Public Safety Comes First in Energy Pinch," *Gettysburg Times*, March 6, 1975, 9.

Page 221   **"It doesn't bother me one bit.":** Ibid.

Page 222   **"who govern for the greater good,:** "Criteria and Eligibility," John F. Kennedy Presidential Library and Museum, https://www.jfklibrary.org/Events-and-Awards/Profile-in-Courage-Award/Criteria-and-Eligibility.aspx.

Page 222   **"My father's heroes,":** John F. Kennedy, *Profiles in Courage* (HarperCollins, 2003), 10.

Page 223   **"We astronauts were good;:** Collins, *Carrying the Fire*, 15.

Page 223   **"My buddies were getting shot at,:** Sington, *Shadow of the Moon*.

Page 223   **"A lot of our friends were flying:** Ibid.

Page 226 **If you do follow your bliss,:** Joseph Campbell with Bill Moyers, *The Power of Myth* (Anchor Books, New York, 1988), 113.

Page 226 **The only limits for us are:** "An Insider's View of Space Exploration Featuring Apollo Astronaut Capt. Alan Bean," YouTube video, 1:00:28, filmed October 12, 2010, posted by the University of Delaware, September 11, 2012, https://www.youtube.com /watch?v=QXavf80lvd0.

Page 227 **There's nobody on this earth like you.:** Ibid.

Page 228 **"I think if you do:** Sington, *Shadow of the Moon.*

Page 229 **"After the flight of Apollo 11:** Ibid.

Page 230 **For, in the final analysis, our:** "John F. Kennedy Speeches: Commencement Address at American University, Washington, D.C., June 10, 1963," John F. Kennedy Presidential Library and Museum, https://www.jfklibrary.org/Research/Research -Aids/JFK-Speeches/American-University_19630610.aspx.

Index

Page numbers in *italics* indicate photographs.

Index

Index

Index

About the Author

Basil Hero is an award-winning former investigative reporter with NBC News television stations. From childhood, and throughout his career as a media entrepreneur and nonprofit executive, Basil has maintained a lifetime fascination with space exploration and the men who went to the moon. He lives in New York City with his wife and two daughters.